Conceptualising Demand

This book addresses fundamental questions about the very idea of demand: how is it constituted, how does it change and how might it be steered?

Conceptualising Demand focuses on five core propositions: that demand is derived from social practices; that it is made and not simply met; that it is materially embedded; that it is temporally unfolding; and that it is modulated through many forms of policy and governance. In working through these claims, the book weaves concepts from the sociology of consumption, science and technology studies, policy analyses and social theories of practice together with empirical cases and new research into such topics as the rise of refrigerated foods, the emergence of online shopping and the transformation of energy demanding services.

This innovative book takes a fresh look at the very idea of demand, a concept that is often taken for granted, but that is vital for scholars and students of energy, mobility, climate change and consumption, and anyone interested in the subject.

Jenny Rinkinen is a researcher in the Consumer Society Research Centre at the University of Helsinki, Finland.

Elizabeth Shove is a Professor of Sociology at Lancaster University, UK, and was the PI of the DEMAND Research Centre.

Greg Marsden is a Professor of Transport Governance at the Institute for Transport Studies at the University of Leeds, UK.

"This is an agenda-setting book. Combining big ideas with telling examples, it shows how we need to follow the practices of daily life if we want to understand the growth of demand. Vital reading for anyone hoping to come to grips with our dangerously growing appetite for energy and mobility."

— *Frank Trentmann, History, Birkbeck College, UK.*

"This book is a response to the urgency of low carbon transitions. It argues for approaches that go beyond energy efficiency and that explicitly focus on how demand is shaped — and can be reduced. The authors sum up insights from decades of research in a way that is accessible and inspiring for new readers and for those already familiar with social practice theory and governance."

— *Inge Røpke, Ecological Economics, Aalborg University, Copenhagen, Denmark.*

"*Conceptualising Demand* provides a coherent account of the social, material and historical foundations of demand, a concept which has become central to many areas of research and policy. A must-read for anyone trying to change demand in the energy or mobility sectors."

— *Yolande Strengers, Emerging Technologies Research Lab, Monash University, Melbourne, Australia.*

Conceptualising Demand
A Distinctive Approach to
Consumption and Practice

**Jenny Rinkinen, Elizabeth Shove
and Greg Marsden**

Routledge
Taylor & Francis Group

LONDON AND NEW YORK

earthscan
from Routledge

First published 2021
by Routledge
4 Park Square, Milton Park, Abingdon, Oxon OX14 4RN

and by Routledge
605 Third Avenue, New York, NY 10017

Routledge is an imprint of the Taylor & Francis Group, an informa business

British Library Cataloguing-in-Publication Data
A catalogue record for this book is available from the British Library

Library of Congress Cataloging-in-Publication Data
Names: Rinkinen, Jenny, author. | Shove, Elizabeth, 1959- author. | Marsden, Greg, author.
Title: Conceptualising demand : a distinctive approach to consumption and practice / Jenny Rinkinen, Elizabeth Shove and Greg Marsden.
Description: Abingdon, Oxon ; New York, NY : Routledge, 2021. | Includes bibliographical references and index.
Identifiers: LCCN 2020006158 (print) | LCCN 2020006159 (ebook) | ISBN 9780367465018 (hbk) | ISBN 9780367465025 (pbk) | ISBN 9781003029113 (ebk)
Subjects: LCSH: Supply and demand. | Consumption (Economics)
Classification: LCC HC79.C6 R545 2020 (print) | LCC HC79.C6 (ebook) | DDC 338.5/212–dc23
LC record available at https://lccn.loc.gov/2020006158
LC ebook record available at https://lccn.loc.gov/2020006159

ISBN: 978-0-367-46501-8 (hbk)
ISBN: 978-0-367-46502-5 (pbk)
ISBN: 978-1-003-02911-3 (ebk)

Typeset in Bembo
by Swales & Willis, Exeter, Devon, UK

Contents

Acknowledgements

This book is one of the outcomes of the DEMAND Centre. We would like to thank everyone who has been involved and contributed to the work of the centre over the years.

This work was supported by the DEMAND: Dynamics of Energy, Mobility and Demand Research Centre funded by the Engineering and Physical Sciences Research Council [grant number EP/K011723/1] as part of the Research Council's UK Energy Programme and by EDF as part of the R&D ECLEER Programme. For more on the DEMAND Centre see: www.demand.ac.uk.

1 Introduction

Most dictionaries define demand as an insistent and peremptory request, made as of right. In economics and in policy, demand refers to the need for goods or services often provided by the market. Generally, this is as far as it goes. In academic debates and in everyday life, demand is simply treated as something that exists, and as something that explains and also underpins trends in what people buy. Despite repeated reference to this term not only in economics, but also in studies of consumption, policy, innovation and technology, the concept of demand has escaped detailed and explicit critical attention.

Many discussions come close. For example, claims about the qualities and characteristics of consumer society make much of the significance and meaning of goods and services in modern social life (Baudrillard, 2016) but leave questions about what consumption is for largely untouched. In economics, demand usually figures as an abstract concept, thought of as the logical partner to supply but rarely considered beyond that. In innovation studies, interest focuses on demand as a stimulus for novelty but again careful analysis of the concept is usually missing (Godin and Lane, 2013). In different areas of policy, rising demand for energy, for health services, for transport or for resources like water (Vörösmarty et al., 2000) is interpreted as an expression of consumers' needs and desires and their willingness to pay for goods and services. Similar understandings inform and underpin welfare economics and the huge machinery of government decision-making. Together, these traditions and conventions combine to form a dominant discourse, built into economic theories, capitalist world views and contemporary accounts of how society works. Across the board, demand seems so obvious, so ubiquitous and so deeply embedded in current thinking that the idea of asking how it is constituted, how it changes and how it might be steered will strike many as

being, at best, odd. Some might conclude, right away, that this is a fool's errand. We think otherwise.

Paying attention to the foundations of demand means paying attention to issues that are sidestepped in mainstream debates, but that are also fundamental to them. For example, many economic theories suppose that demand reflects prices and individual preferences. Such approaches capture just a small part of why demand changes and are largely blind to longer-term historical trajectories, or to social institutions, infrastructures and practices. As Shove and Walker (2014) argue, and as indicated within energy and transport related research and policy (Marsden et al., 2018), there is a parallel tendency to focus on matters of efficiency or security of supply rather than on basic questions about what energy and mobility are for. This results in partial but hugely influential representations of demand that inform what are often short-term, and very limited ideas about what policy makers can do to tackle challenges like those of radically reducing carbon emissions. In taking the foundations of demand for granted, and in taking them out of the equation, debates about what can be done revolve around a handful of alternative responses. In the context of consumption, carbon emissions and climate change this results in a lopsided discussion about whether technological solutions can fix the problem or whether individuals also have to cut back or deny themselves goods and services that have become synonymous with normal ways of life. Framing the problem in these terms bypasses primary questions about how expectations of normality form and change. This neglect has important consequences for policy agendas. One is that in the energy and mobility sectors, policy ambitions shrink to fit what are taken to be non-negotiable interpretations of normality. Proposed solutions consequently focus on opportunities for efficiency or 'behaviour change' within a narrow frame, leaving big questions about how present and future demands evolve out of view. This is not just an unfortunate omission. In routinely taking current ways of living as their point of reference, energy and environmental policies inadvertently *reproduce* patterns of consumption that are themselves part of the problem.

One reason for writing this book, and one reason for focusing on energy and transport is that climate change represents what is probably the most pressing social and political challenge of the day. If fossil fuel consumption is to fall at the rate and on the scale required, questions about the foundations and the future of demand will have to take centre stage. Amongst much else, this requires better understanding of

the qualities, origins and histories of demand, and of how these emerge and change over time.

In pursuing this agenda, we challenge what have become familiar responses in climate change research and policy. To be more specific, we take issue with the ways in which strategies of efficiency and decarbonisation have been divorced from issues of demand. In many countries, including the UK, efforts to reduce energy consumption and promote the transition to a lower carbon economy depend on decarbonising energy supply by making more and better use of renewable energy, or on delivering the same or more service but for less energy, for instance through the development of more efficient cars, heating and cooling systems and appliances.

It is widely recognised that the introduction of lower carbon technologies will have implications for 'markets, infrastructures, institutions, social practices and cultural norms' (Jenkins and Hobson, 2018: 2). Linked to this, there is an extensive literature that emphasises the 'sociotechnical' character of sustainable transitions and that considers the various roles that users play in these processes (Schot et al., 2016). Despite interest in the practicalities of adoption and uptake those who write about energy transitions rarely consider how infrastructures and technologies (including efficient ones) configure and transform social practices and the forms of energy demand that these engender and sustain. Again the basic dynamics of demand are largely taken for granted.

Instead, and as evident in recent 'clean growth' strategies, demand – here meaning the amount of energy that society needs – is treated as given. Estimates and models of future demand work in fairly predictable ways: expected changes in population, estimates of GDP and anticipated technological trajectories such as the uptake of electric vehicles are factored into equations that assume that current standards of living and present energy-demanding practices will persist unchanged (DfT, 2018b; National Grid, 2017). With these understandings in place, the question is how to meet these present and future needs whilst achieving targets for carbon reduction (Defra, 2017; UK Government, 2017b).

This is not to suggest that policy makers and researchers disregard or take no account of demand. Far from it. As indicated above, judgements about how much energy society needs are woven into forms of policy analysis, future investment, energy modelling and more. Rather, the point and also the purpose of this book is to articulate and challenge the tacit theories and understandings on which these assumptions depend.

From this perspective, the tendency to conceptualise demand as an expression and outcome of consumer choice or as a consequence of technological efficiency is a significantly limiting feature. It is so in that discussions of demand then revolve around matters of individual behaviour or focus on peoples' values, preference and abilities to pay. This is such a familiar approach that commentators rarely give the underlying logic a second thought. For example, it is at first difficult to disagree with those who point to the many benefits of a circular economy, involving forms of sharing, recycling and reuse. Following this line of reasoning it makes sense to focus on persuading people to play their part, for instance, by recycling clothing. The drawback of such an approach is that it locates both the problem and the solution in the hands of the consumer. Very different responses would be required if trends in resource use were attributed to the massive increase in the global production and distribution of clothing in the first place (Trentmann, 2016). This is not the path that is usually taken. Instead, dominant interpretations of demand generate two also dominant types of intervention, namely raising consumer awareness, and increasing efficiency.

Thinking of demand as a consequence of the sum total of individual choices implies that it can be reduced through changes in market conditions (e.g. price), by raising awareness of the consequences or by providing information that enables people to make 'better' choices for themselves. Amongst much else, arguments like these justify efforts to engage consumers in so-called 'smart behaviour', exemplified in the energy sector by the rolling out of 'smart meters' that allow people to see their gas and electricity consumption in real time (Strengers and Hazas, 2019).

Alongside these initiatives, demand reduction policies also focus on developing and promoting more efficient cars, fridges, tumble dryers and so forth. In some respects this strategy makes sense: there are energy savings to be made. However, there are also unintended consequences, some of which are positively counter-productive. Discussions of the downsides of efficiency tend to focus on the problem of rebound. In brief the concept of rebound refers to the possibility that the money saved through buying and using a more efficient appliance might be spent in ways that increase energy consumption in another area of daily life: for instance enabling more travel, or flights abroad (Sorrell, 2007; Marsden, 2019). However, a much more basic problem is that efficiency programmes legitimise and even promote the use of resource intensive appliances. It is all very well to develop a tumble dryer that uses less energy than another existing model, but why is it

that tumble dryers are, in any case in use? Discourses of efficiency compare like with like meaning that there is never any discussion of how efficient a washing line might be, compared with a mechanical dryer. Similarly, the need for a fridge or fridge freezer (or two) is simply not part of the efficiency discussion. Nor is an understanding how global food systems have come to depend on unbroken chains of refrigeration.

In this book we argue that as long as policy makers, researchers and practitioners think of demand as an expression of individual consumer choice, as an outcome of the self-evident need for specific goods and services, or a consequence of their design they are unlikely to do more than scratch the surface. In essence, our argument is that demand depends on the range of practices enacted in society, and on the technologies, infrastructures and institutions on which these depend. All practices generate and are linked to some kind of demand whether that be for a vehicle and road infrastructure to get to work or for electricity and a kettle for making tea. From this point of view, the extent, and also the timing of energy consumption is a result of how demands for different forms of service (thermal comfort, convenience, hygiene) are built, and how these change.

In developing an account of the social basis of demand we have two ambitions. One is to bring what are usually sunken questions of history and change out into the open. The second is to demonstrate the policy relevance of engaging with the constitution of demand in this more fundamental sense.

Demand – extending the agenda

In the rest of this book we work through the theoretical and practical implications of conceptualising demand not as an expression of individual choice, but as an outcome of shifting, historically situated complexes of social practice (Shove et al., 2012).

Such a focus on the very foundations of demand is, we argue, urgently required given claims about the limits to growth (Meadows et al., 1972) in terms of population increase, agricultural production, non-renewable resource depletion, industrial output, pollution and carbon emissions. Rather than expecting that these problems can be fixed by means of efficiency improvements and other large-scale engineering, such as GMO agriculture, geoengineering or artificial intelligence, our aim is to explore opportunities for reducing the scale of these challenges not through new forms of supply, but by actively reconfiguring demand.

This is an unusual and also challenging approach. There are many good reasons why demand is so often taken for granted, and why it is so often excluded from serious debate. As we have already mentioned, dominant ideologies of neoliberal consumer society, of the state and of capitalism itself deny that demand is any other than an expression of consumer choice. In addition and from the Brundtland report onwards, there has been a consistent emphasis on wealth, growth and prosperity:

> Many essential human needs can be met only through goods and services provided by industry, and the shift to sustainable development must be powered by continuing flow of wealth from industry.
>
> (Brundtland et al., 1987: 16)

We are consequently writing in a context in which the focus on choice and growth is such that the only viable courses of action appear to be those that fit within this frame. This is consistent with ideas about the scope of legitimate government intervention, and with the ambition of 'helping people make better choices for themselves' (Shove, 2010). From this point of view, it is quite appropriate for climate change policy makers to encourage and promote efficiency, but quite out of order to advocate laundering less, using a clothes line or doing without a fridge. Put differently, the focus is on methods of meeting demand as it is currently constituted.

However, if we take even a short-term historical view, it is obvious that policy makers also have a hand in *making* demand. For example, many areas of daily life and many of the forms of demand that follow are directly and indirectly affected by guidelines and regulations including those relating to health and safety, efficiency standards, land and resource use, and so forth. Similarly, large-scale infrastructures and systems of provision, from electricity and gas grids through to road and rail networks, are designed and sized to enable certain ways of life. In other words, issues of demand are, implicitly at least, at the heart of many critical public policy dilemmas, such as determining appropriate levels of spending on health care, on public health provision, policing or education. In all these contexts and more, the challenges of managing demand are not fixed or stable.

Developments in some areas of everyday life generate forms of scarcity (as when more land is needed for housing) and conflict, as when air pollution increases with urban density and traffic. The practicalities and politics of response are correspondingly complicated and in both the public and the private sector there are repeated failures to grasp

changes in demand. This takes many different forms. Often there are mismatches between provision and practice, as when rail franchises are tied to infrastructures that cannot keep pace with changing patterns of commuting and travel, or when hospital buildings are designed to accommodate forms of treatment that are swiftly outdated. As the ensuing struggles indicate, the response is almost always one of adapting and modifying provision in order to meet demand, whenever it occurs and however extreme it might be. At the same time, it is obvious that definitions and meanings of appropriate service, and related systems of provision are themselves in flux. For example, some electricity utilities now sell services and not just power. Similarly, some manufacturers are deliberately seeking to extend the life cycle of their products, making use of recycled materials, 'upscaling' and aiming for sustained rather than increased production. In both these cases, providers are involved in making new and different forms of demand.

Despite the rhetoric, and whatever the goals that producers and policy makers might have, it is increasingly obvious that demand *does not* simply exist, ready-made. As indicated above, it is actively constituted within and outside markets, and in settings that are multiple and varied. This is an important insight but it is also just the start. In the chapters that follow we see demand as an outcome of social practices. Because of this, we take a broader-than-usual view of what is at stake. This enables us to develop an also broader-than-usual account of how demand might be steered and shaped, and of how significant challenges like those of radical carbon reduction might be approached.

We have already referred to social practices, as distinct from consumer choice and behaviour, but what do we mean when we say that demand is an outcome of practice? When we refer to social practices, we mean more than what people do at home, at work or in moving around. As outlined below, when we talk of practice, we do so in a way that is informed by, and that carries with it, a distinctive set of theoretical commitments.

In social theory, practices, as defined by Giddens (1984), exist across space and time and have lives that extend beyond any one person and beyond any one moment of enactment. As Reckwitz puts it, 'The single individual – as a bodily and mental agent – then acts as the "carrier" (Träger) of a practice' (Reckwitz, 2002: 250). This is an important theoretical distinction in that it locates practices, and not individuals, as the central topics of conceptualisation and analysis.

For Giddens, and for us, studies of practice overcome classic distinctions between agency and structure, emphasising 'the essential recursiveness of social life, as constituted in social practices' (Giddens,

1979: 5). To put it more concretely, social practices are formed and reformed through historical developments and interconnections, they are shaped by and shaping of societies, and they are entwined with configurations of space and time, as well as with infrastructures, materials and institutions (Shove, 2003; Shove et al., 2012). Equally, it is only because they are more or less persistently, and more or less consistently enacted that they endure and change. These ideas, along with the related contention that practices are *the* 'site' of the social (Schatzki, 2002), and that societies are none other than an immensely complex 'plenum' of practices have inspired a range of social theoretical debates about the constitution and the transformation of this nexus (Hui et al., 2017). These discussions underlie and inform key features of our approach to demand in quite specific ways.

Building on the core tenets of practice theory, we claim that demand is an outcome of the social, infrastructural and institutional constitution of society and that resources such as energy are consumed and transformed in accomplishing a huge range of social practices including those involved in heating, commuting, laundering, cooling and so forth. In each case, the details are not pre-given: what practices 'require' changes over time, at different rates and in different ways. The extent and timing of energy demands is thus a consequence of these many and multiple arrangements. It follows that increases, or decreases, in demand depend above all, on the combined trajectories of the sum total of the practices involved. Our view of demand is thus pluralistic and multidimensional. We recognise that demands are multiple and that they are made and met within and outside of markets. Demand for goods, services, and resources such as energy combine in ways that relate to the demand for what we describe as meta-services like mobility and comfort. Conceptualising demand as an outcome of complexes of social practice underlines the point that it is constantly and actively reproduced: it does not simply exist as something to be met, nor does it remain unchanged.

These starting points open the way for further analysis of the material and also the temporal constitution of social practice, and thus the shaping of demand (Schatzki, 2011). In building on these ideas, and in weaving them together, this book is organised around five propositions:

1. Demand is derived from practices
2. Demand is made, not simply met
3. Demand is materially embedded
4. Demand is temporally unfolding

5. Demand is modified and modulated, deliberately or not, via many forms of policy and governance

The chapters that follow work through these propositions in turn, doing so as a means of elaborating on the characteristics and the dynamics of demand for energy and mobility, in particular.

In developing this approach we draw from, and also challenge and critique a range of disciplines and traditions. What we have to say is anchored in an understanding of social practice (taking this to be the basis of demand), but in opening up new lines of enquiry we open the way for an influx of fresh ideas and intellectual resources from policy studies, science and technology studies and more. At the same time, there is no denying that ours is a challenging approach: it challenges the individualistic and also abstracted conceptualisations of demand on which much contemporary policy depends and it questions the kinds of responses that follow, rejecting simple narratives of efficiency, and favouring more concerted interventions that engage with the vey basis of demand. Above all, it takes issue with the overly limited scope and ambition of what counts as energy, or as transport, or as climate change policy. In taking these ideas forward, we show how questions of demand – what demand is, how it changes – can be reframed, and we identify the implications of such reframing for public policy.

Before getting underway it is useful to say more about the choices we have made in bounding the topics we discuss, and the scope of our analysis.

First, we recognise that not all forms of energy or transport demand are problematic, but we share the view that resulting and aggregate levels of related carbon emissions are so now and will be even more so in the future. In thinking about which forms of demand matter, and how, commentators distinguish between 'good' and 'bad' or 'wasteful' and 'useful' types of energy consumption, often linking these definitions to interpretations of 'necessities' and 'luxuries' that are in turn associated with broader understandings of well-being at different scales (Büchs and Koch, 2019). Some discussions of sustainability start from this point, for example setting out minimum and maximum standards for consumption (Di Giulio and Fuchs, 2014).

We take a different route. Rather than making our own judgements about good and bad, our aim is to detail the connections between social practices, the forms of demand (for goods and resources and also services like mobility; care, etc.), and the organisation of provision that follow. This makes sense in that we do not take

practices, and related interpretations of wellbeing, to exist somehow independent of systems of provision, or of the materials and technologies involved. What counts as a necessity and what is seen as waste is *part of* the dynamics of practice, and as such part of what needs to be studied and observed.

Second, and as outlined above, we focus on examples relating to energy and mobility because we believe, in very general terms, that both are critical for carbon emissions and climate change. Yet there are further decisions to be made about the scope of our enquiry. The demand for energy (meaning resources like oil, electricity or gas) does not take place in a vacuum, but it is closely linked and intertwined with demand for related goods and services. In practice it is not energy, itself, that is demanded – rather it is the services that energy makes possible. Further, the services that matter are those that enable or constitute one or more social practices.

As Morley (2018) explains, transport systems, which depend on the actions and strategies of many different organisations, provide what is best conceptualised as a background or enabling condition for a range of other very varied practices: commuting to work, taking a holiday, going shopping etc. Similarly, the technologies and systems that constitute what now counts as 'comfort' (including the fabric of a house, clothing and heating systems) enable people to do all sorts of things: to read, sleep, cook and so on. Morley describes mobility and comfort as 'meta services', using this term to distinguish these encompassing forms from other more specialised relations, for instance between the energy used to power a specific device, and the service that device provides. Meta-services like comfort are never fixed entities: rather they have histories, they depend on suites of technologies, and they change when their components change. In Morley's words:

> by focusing on meta-services as more encompassing formations of convention, experience and means of provision, in which energy services like lighting and heating are usually only one amongst many components, we can better recognise the dynamic and collectively distributed nature of demand.
>
> (Morley, 2019: 20–21)

Showing how such combinations of technologies and services change *together* represents one means of showing how demand evolves at a societal scale.

This way of thinking puts issues of increase and reduction in a new light. For example, energy efficiency measures reduce the amount of energy consumed whilst maintaining the same level of service, and the same contribution to broader concepts of meta-service. Such measures limit consumption but do not modify demand – not in the fuller sense of reconfiguring what energy is 'for' or the practices on which consumption depends.

Working with this broader-than-usual interpretation of demand enables us to see how the components of meta-services are arranged and rearranged, and to consider the escalation or reduction of resource use. To illustrate, there are different ways of making oneself comfortable: clothing can be part of this endeavour, or the addition of mechanical heating or cooling. Which components (clothing, or heating) figure more or less prominently and how they interact changes historically and with significant consequences for fuel consumption. The approach we take allows us to keep this shuffling in view.

This is not just a matter of recognising that the 'same' practice can be done in more or less resource intensive ways (Heikkurinen, 2018). It is, of course the case that swimming in the sea has different energy implications compared to swimming in a heated pool, and services such as communication, illumination, hygiene, sustenance or nourishment, mobility or transport, shelter or structure, and thermal comfort can all be achieved in a variety of ways. However, a third distinctive feature of our approach is that we resist the tendency to think about comfort, light, shelter etc. as enduring and universal forms of 'need', desire or function that are always present or that must be satisfied in one way or another. For us, the more interesting question is how practices such as office working, cooking or laundering evolve through and along with technologies that are themselves dependent on background infrastructures of energy and power. It is in this sense that demand is quite literally woven into the fabric of daily life. Conceptualising demand in these terms treats resources (including gas and oil, but also things like steel, sugar or coffee) not as unchanging commodities exchanged in markets driven by seemingly abstract political and economic processes, but as integral elements within living systems of practice.

Fourth, and related to this, we go beyond representations of energy demand that are, in effect, fixed on resources, taken in isolation. Rather than being off-limits, questions about the 'ends' of energy services – i.e. what energy resources are used for, and how this changes over time – are central. This plays out in different ways. For example, in the transport sector it is important to understand that something as apparently

stable as peak hour traffic congestion is currently comprised of a very different set of journeys and purposes than was the case a decade ago (Le Vine et al., 2017). Rather than being simply tied to price, or to consumer preference our contention is that trends like these are 'related to social, cultural and collective histories and traditions' linked to 'sociotechnical change and the co-evolution of infrastructures, devices, routines and habits' along with the 'inter-dependent practices of producers, providers, utilities and governments' (Wilhite et al., 2000: 118). From this point of view it is not simply the demand for services or the use of resources that matters but the relationship between them. This argues for forms of analysis that consider the dynamics and also the economics not of fuel use, but of meta-services like comfort, illumination or mobility.

Finally, a word on sectors other than energy and transport. Many of the arguments we make and the positions we develop would apply as well to discussions about the demand for water, or for education, policing or health care. At the same time, there are revealing and instructive differences in how demand is conceptualised. For example, in the transport sector, demand is often said to be derived from what people do, an interpretation that fits well with the approach we take, but that hardly ever figures in discussions of energy use in buildings. Similarly, although there is no obvious equivalent in the energy world, in public health, the concept of an obesogenic environment, that is an environment that favours practices that contribute to obesity, has informed interventions that focus on infrastructures and conditions that might enable healthier patterns of diet and exercise. As these more adventurous responses remind us, social practices and related patterns of demand are not fixed for all time. Nor are they immune from policy influence. Whether they are aware of it or not, policy makers are constantly intervening in ways that shape the very foundations of demand. What is missing, and what we hope to provide, is an account of the processes involved. It is for this reason that we write about how demand is constituted, how it changes and how it might be shaped and steered.

Structure of the book

The next four chapters develop and explore the central propositions on which our distinctive approach is based.

Demand is an outcome of practice and is made and not simply met

Chapter 2 starts by describing the theoretical roots of currently dominant ways of thinking about demand and closely related notions of

need, want and desire. Having discussed the status of demand within classical economics, and in social theories of consumption we say more about what is involved in viewing it as an outcome of social practice. As we explain, the contention that demand happens through and as part of practices calls for a more detailed account of relations between materiality, practice and demand.

Demand is materially embedded

Chapter 3 takes up this challenge and explores the proposition that demand for goods and services is rooted in the material arrangements on which social practices depend, and that are in turn part of shaping what people do. In this chapter our aim is to show how infrastructures and appliances figure in the conduct of practice, in the constitution of meta-services and in how social life is held together. We use two empirical examples, the diffusion of fridge freezers and the development of online shopping, to show how sets of practices, the material arrangements on which they depend and the forms of demand that follow link and change together.

Demand is temporally unfolding

Chapter 4 is about how demand changes and varies over time. Rather than dealing with long-term changes in the constitution of practices this chapter deals with fluctuations in demand during the day and over the year. The suggestion that the timing of demand is a consequence of how social practices are synchronised and sequenced has important consequences for the energy and mobility sectors and for efforts to reduce peak loads, limit congestion and make better use of renewable forms of energy supply. The suggestion that forms of provision and supply are part and parcel of shaping and also responding to patterns of time and timing in what people do is an important part of our analysis, as is understanding of flexibility as an outcome of how complexes of social practice hang together. These ideas inspire new ways of thinking about the timing of supply and demand in energy and transport systems, and about where opportunities for intervention might lie.

Demand is modified and modulated, deliberately or not, via many forms of policy and governance

Chapter 5 investigates some of the ways in which policy influences the practices and social arrangements on which energy and mobility

demand depends. We use the case of transport policy to show how dominant paradigms and methods of framing problems and forecasts underpin forms of investment and 'solutions' that inadvertently sustain increases, rather than reductions in travel demand. We also argue that other approaches are possible. If taken to heart, the propositions set out in this book lay the foundations for policy interventions that are capable of steering demand, and of doing so in ways that are coherent and compatible with swift and significant responses to the challenges of climate change and carbon reduction.

Rather than summarising the book as a whole the final chapter highlights what is distinctive about our approach. In bringing the threads together, we explain why it is so critical that those who take demand for granted, who take it to be the logical partner to supply, or an expression of consumers' needs and desires should think again about the assumptions they make. The chapter does not provide easy policy solutions but it shows that conceptualising demand in the way that we do can help identify opportunities that are simply not on the radar in energy and transport research and policy, not as it is currently framed and formulated. This is important in that radically new concepts and strategies are urgently required if there is to be any hope of moving towards a low carbon society. This book gives a sense of what this might entail.

2   Constituting demand

This chapter is about how demand is made in two different senses. We begin by considering how demand is conceptualised and in that sense constituted within economics, market studies and consumer research. Having described different ways of thinking about this concept and about related terms like consumption, desire and need, we introduce and develop the suggestion that the demand for goods and services is derived from social practices, and that it is quite literally constituted and changed as practices evolve across space and time. As we explain, the proposition that goods and services are always and inevitably embedded in the many and varied practices of which society is composed has profound implications for understanding both supply and demand. It is also critical for how we think about the roles that different sorts of materials play in daily life, and in the timing as well as the extent of demand. We have much more to say about the implications of these ideas but as promised, we start by taking stock of how demand figures in different fields. The next two sections sketch alternative interpretations of demand, showing how they vary and what they have in common.

Demand, supply and the laws of the market

It might seem bizarre to suggest that classical economic theory has failed to deal with demand. After all, demand has been a key concept ever since Adam Smith (1977) argued that the relation between supply and demand acts as an 'invisible hand' in guiding markets. The idea that this relation is one of the primary economic forces in society has several related features. One is that consumer goods and services are thought of as commodities that are produced, distributed and supplied in response to consumer demand. Second is a tendency to suppose that demand and supply are subject to various generic and

universal laws of the market. Ideas about the functioning of price mechanisms provide a good illustration of how these law-like relations are thought to operate. A first necessary step is to suppose that levels of demand are founded on individual preferences. For example, a person's preference for ice cream is measured by how much they would be willing to pay for one: perhaps £3. Then, what would they pay for a second ice cream? Given that people generally enjoy their first ice cream more than their second this might be £2. After two ice creams the imaginary consumer is getting a bit full, meaning that he or she might only be willing to pay 50 pence for the third, and so on. In theory, if everyone was willing and able to declare their preference for initial and subsequent ice creams then we would have access to every individual demand curve which we could add together to develop a societal demand curve.

As this simple thought experiment shows there is no history and no future in this representation of demand: it is a snapshot of preferences holding everything else equal. There is no mechanism for understanding changing demand over time (rather than as a result of price) other than assuming that preferences simply shift. In other words, there is no model for understanding what it is that results in preferences changing, or in the production and structuring of the 'options' on offer. Instead, demand is thought to be tied to supply through a 'self-regulatory' mechanism linked to a 'law' of market equilibrium meaning that products demanded at a price are equalled by products supplied at that price (McConnell et al., 2009). This basic understanding leads commentators to identify what are described as the 'drivers' of demand, of which price is one. In essence, the assumption is that if prices rise, people will cut back on how much they consume (demand will reduce), or buy somewhat more if costs go down (demand increases). Other drivers can also come into play. These might be identified by, for example, measuring how patterns of ice cream consumption vary with outdoor temperature. The method of isolating and analysing the effects of individual factors, one by one, results in similarly ahistorical representations of demand and how it changes. To continue, understandings of weather as a driver of ice cream demand would be indifferent to the seasonal organisation of daily life, to the large number of day trips to the coast or to a park on hot days or to the also seasonal lives of ice cream sellers. Put simply, the wider social practices which change and on which changes in demand depend are absent from talk of preferences and drivers.

Patterns of ice cream consumption are not the same as patterns of energy demand and as is obvious, not all goods and services are alike.

In response, economists have introduced additional terms to explain what are seen as discrepancies and deviations between supply–demand relations. The notion of price elasticity is, for instance, used to account for the fact that when prices rise, consumers are more willing, or able, to forego some items and thus reduce some forms of demand than others. Inelastic forms of demand are those that tend to persist despite fluctuations in price.

The notion of price elasticity is an essential part of economic theory and people working in this tradition treat it as a 'fact', or, more accurately, as a correction factor that explains why markets work as they do. Elasticity consequently figures within and as part of what are essentially abstract accounts of supply–demand relationships and market processes. In these contexts, ideas about the relative elasticity of demand are again discussed without reference to history or context. As Torriti explains, economic theories have no account of how specific elasticities come to be what they are, or why some items are more elastic than others (Torriti, 2019). Fouquet also points out that economists have been measuring the price elasticity of energy demand for decades, notwithstanding significant changes in the uses to which energy is put, and in the intensities of different fuels per units of input (Fouquet, 2014). Such considerations are out of scope because economics operates in what Callon and Muniesa describe as a realm of 'pure calculation', developing and working with models and variables that constitute a kind of parallel universe at one remove from the complexities of actual economic exchange, and from the historically situated practices involved (Callon and Muniesa, 2005).

Insofar as the demand for specific goods has an origin or a cause this is implicitly anchored in consumers' needs, preferences or desires and their ability to pay. Although all of these considerations develop and change over time, discussions of demand tend to focus on the *amount* and *type* of resources/services people consume under specific conditions. More than that, they tend to do so without enquiring into the reasons behind these arrangements, and without considering the possibility that forms of supply might be implicated in making and not simply meeting demand.

Instead, producers and providers are thought to have the more limited role of identifying and meeting consumers' needs, and of designing and organising systems of provision so that supply matches demand at a price which suits both suppliers and consumers. This kind of reasoning treats demand as something that varies in response to a handful of drivers and trends, for instance in urbanisation, lifestyle, population and economic growth (Asif and Muneer, 2007), but

that has an existence of its own. Put differently, economists and the policy makers they inform rarely grasp the specific dynamic processes through which demand grows or shrinks in response to supply, nor do they take account of the fact that forecasts and estimates of future demand actively contribute to these developments.

We discuss the performative role of methods of estimating present and future demand in Chapter 5, and in Chapter 3 we write about how infrastructures and systems of provision are implicated in shaping and constituting the very practices that they enable. For now it is enough to notice that economic theories tend to 1) treat supply and demand as abstract concepts, 2) suppose that demand exists independent of the ways in which it is met, 3) equate consumption – that is the acquisition and use of goods and services with demand and 4) suggest that there are identifiable and also generic drivers of demand, including price and consumer preference.

In the next section we take stock of social scientific interpretations of demand that focus on the social and cultural meaning of consumption, and the constitution of needs, wants and desires.

Demand is a consequence of consumers' needs, wants and desires

In market studies, as in much consumer research, 'consumption' is often equated with the decision or commitment to buy. In this context, consumption is taken to be an expression of individual choice, based on the ambition of satisfying needs and meeting desires and wants, limited or enabled by a willingness or ability to pay. In such representations, demand (here meaning the act of purchasing) depends on more than price since the consumer 'experience' includes aspects of imagination, emotion and evaluation (Holbrook and Hirschman, 1982).

Despite these additional complexities, there is considerable overlap with economic theories of demand, including a shared focus on the decisions and actions of more or less rational individual consumers (McCracken, 1990). For example, approaches stemming from theories of planned behaviour (Ajzen, 1991) suppose that consumer choices are driven by individual attitudes, and that changing patterns of consumption is in essence a matter of changing peoples' environmental values (for a critique see Shove, 2010).

The idea that individuals act in ways that reflect and that are driven by their beliefs and attitudes is a staple feature of well-established and essentially psychological theories of behaviour. It is also a view that is widely shared in policy and in popular discourse as well. Thaler and

Sunstein's much cited ideas about what policy makers can do to 'nudge' people in the right direction and to help them make better choices for themselves exemplify this tradition (Thaler and Sunstein, 2009). Programmes and policies that are designed to reduce energy demand, and that are informed by this approach, focus on raising consumer awareness, modifying price signals and reducing what are known as 'non-technical' barriers. The basic idea is that measures like these will modify beliefs and knowledge with the result that consumers will then act appropriately: whether by setting the thermostat back a notch, purchasing an electric vehicle or buying a heat pump. As set out in a recent report entitled 'Behaviour Change, Public Engagement and Net Zero', 'Policy is needed which supports consumers to take specific, high-impact actions by lowering barriers and enabling more informed choices' (Carmichael, 2019: 18).

Within the psychological and policy literature there is some discussion of factors that complicate what are otherwise linear accounts of the drivers of behaviour change and the determinants of demand. These include things like habits that are stubbornly hard to shift, and that often persist despite targeted advice and information (Darnton et al., 2011). Similarly, ideas about the so-called value-action gap are introduced to help understand why people fail to follow through on their environmental commitments (Barr, 2006). In both cases, contextual and situational considerations are patched in to theories that are, in essence, focused on the project of explaining individual action. In addition, and as with the economic theories and approaches discussed above, accounts like these take no note of longer term historical trends. In terms of policy making, the focus is on nudging choices made in the present and in relation to a limited range of actions that are 'normal, easy and in alignment with other day-to-day concerns (e.g., household budgets and social relations)' (Carmichael, 2019: 19).

Although there is often some reference to broader themes of 'public acceptance' those who write about behaviour change rarely explore differences of class and culture. However, these are important topics, especially amongst sociologists of consumption and culture. For Bourdieu, goods and services – including those relating to transport and energy demand – help define and reproduce social distinctions and social groups (Bourdieu, 1984). Developing these ideas, Holt suggests that purchasing decisions are indicative of broader notions of lifestyle and identity alongside attitudes and values (Holt, 1997). This suggests that patterns of demand are unevenly and recursively linked to social position and status. From this point of view, the constitution of value and meaning are part of the story.

These are important matters for anthropologists inspired by Marx's theory of value and for those who argue that consumption is a form of symbolic action (Gell, 1986). For authors such as Appadurai (1994) commodities embody social as well as economic value. Understanding how objects acquire and lose meaning and value is consequently part of understanding how specific forms of consumption come to be seen as 'normal' or 'easy'. From this point of view, relations between supply and demand are not simply defined by the goal of meeting established needs and wants. Instead, the symbolic meanings of consumption, often related to the ongoing formation of identity within and as part of consumer culture (Belk, 1988), take centre stage.

The wider cultural turn in the social sciences (Geertz, 1973) has generated further interest in the characteristics of consumer society (Arnould and Thompson, 2005) and in the social and cultural qualities of 'the market' and of market exchanges in general (Callon, 1998; Miller, 2002; Slater, 2002). This has led to a reinterpretation of the economic sphere. Rather than treating this as a separate realm, uncomplicated by social relations of difference, meaning and valuation, those who take a 'pragmatic' approach to market studies write about how relations of exchange, including consumption and provision and, by implication, supply and demand are constituted in practice.

Some of this discussion has to do with the forms of 'entanglement' and 'disentanglement' involved in establishing and positioning things as commodities: that is as objects that circulate within markets. There has been extensive discussion about the 'work' involved in making markets that resemble and that are informed by the abstract constructions of economic theory. For example, some contend that effective market exchange is only possible when goods are stripped of (certain) meanings, and when they become anonymised commodities through the potentially alienating processes of buying and selling. Others are interested in the extent to which the commercial value of an object is, or is not, a 'good proxy for its social and cultural value' (Sassatelli, 2007: 10). Again, views differ on how economic value is made and on the forms of cultural and historical shaping involved (Appadurai, 1994).

Often those who write about consumer culture do so in rather broad terms, rarely making the link between the social significance of acquisition and the practicalities of producing and providing specific goods and services. Instead, and insofar as it is considered at all, the relation between demand and supply is conceptualised as something that enables 'consumer culture' and the circulation of lifestyle-related goods (Featherstone, 2007). To some extent this reflects a consistent focus on the present at the expense of understanding how patterns of

consumption evolve over time (Warde, 2014). It also reflects an enduring interest in more conspicuous features of consumer culture (Shove and Warde, 2002).

For decades, studies of consumer culture paid relatively little attention to unglamorous and mundane forms of consumption (Gronow and Warde, 2001; Shove, 2003), or to the close-coupled relation between resources, objects and the range of practices they enable. Not surprisingly, consideration of 'ordinary' consumption (Gronow and Warde, 2001), including the consumption of electricity and water, drew attention both to the limits of consumer choice, and to the fact that demand is derived from, and embedded in other activities. Rather than seeing the consumer as someone continuously making deliberate choices, or enacting personal values, writing on inconspicuous consumption emphasised the importance of networks and systems of provision (Warde, 2005), and the routinised nature of much consumption, especially that relating to dwelling, eating and transport.

Growing interest in mundane but globally significant forms of consumption led to renewed interest in processes of use rather than moments of acquisition, and to also relevant research in science and technology studies. For example, analyses of 'domestication processes' through which appliances and technologies become embedded in daily life (Callon, 1984; Silverstone and Hirsch, 1992) have the dual function of showing how consumer goods relate to each other, and to the changing practices of those who use them. To give a specific example, things like fridge freezers clearly have an impact on patterns of shopping and cooking, and on the range of foods available to buy. This is linked to and cannot be understood aside from the ways in which the global food industry is organised (Shove and Southerton, 2000; Hand and Shove, 2007; Rinkinen et al., 2017). These and related lines of enquiry have expanded the scope of consumer studies. For example, authors like Warde (2005) have argued that consumption is part of every practice, and that practices are, by definition, shared, social and distributed across space and time. Amongst much else, these ideas challenge the view that demand arises from individual consumers' desires or that it is an expression of what are taken to be universal, non-negotiated needs.

It is true that people suffer and ultimately die if they are unable to eat or drink fresh water or if they get extremely hot or cold. This has led some authors to conclude that human beings have certain 'basic' needs, for example, for shelter, clean water and thermal comfort but also for education and a non-threatening environment (Walker et al., 2016), and that these underpin at least some demand. Not surprisingly,

there are different views about precisely what people need to consume in order to live well or to participate effectively in society. For example, many suppose that 'basic' needs are universal (Doyal and Gough, 1991). Some agree but claim that there are different, historically specific means of satisfying them. Meanwhile, others deny the existence of universal needs but argue that in any one society and at any one moment, normative judgements are made about what counts as an acceptable standard of living (Day et al., 2016). As Day et al. (2016) show, from this perspective, demand, for instance for things like a TV, a laptop or broadband access, reflects the fact that in some societies and for some social groups these now count as normal and necessary services and resources.

A more subtle point, and one that brings us back to a discussion of social practice, is that goods and services are not important or 'needed' in their own right, but because of what they enable people to do. This argues for tying discussions about needs, desires and demand back to an historical understanding of how related and connected social practices emerge and change, and how these dynamic processes define and constitute interpretations of necessary or desirable goods.

In economics, in consumer studies and in discussions of need and entitlement, fundamental questions about demand – that is about what consumer goods and services are 'for' and how this changes – are rarely at the forefront of analysis. The reasons for this vary depending on the disciplinary tradition involved. For example, in economics, demand is often equated with consumption as revealed by sales data. Alternatively, it is taken to be the unproblematic consequence of trends in consumer culture including the symbolic significance and meanings attached to specific consumer goods. In addition, and as indicated above, some writers think of (some) demand as the necessary consequence of meeting universal human needs, or socially determined expressions of them.

What is absent, and still elusive, is an account of precisely how the 'world of goods' and services relates to the worlds of social practice, and thus of how needs, requirements and desires are formed and changed. In the next part of the chapter we introduce and discuss two related propositions 1) that demand is derived from practices – in other words it is an outcome of the social, infrastructural and institutional ordering of what people do (Shove and Walker, 2014) and 2) that demand is made, not simply met. Defined in this way, demand has multiple histories and it is constantly changing in line with forms of provision and supply, and with the practices on which it depends and from which it is derived.

Proposition 1: demand is an outcome of social practice

Theories of practice have made important contributions not only to the understanding of consumption (Warde, 2005), but also to meanings of service (Shove, 2003), and the role of infrastructures (Shove and Trentmann, 2018) and materials in social processes (Maller and Strengers, 2018). There are various ways of conceptualising social practices but most take them to be recognisable but constantly changing 'entities' reproduced through performances (moments of doing and saying); and involving the active integration of elements such as materials, meanings and competencies (Shove et al., 2012). With these ideas in place, further questions arise about how practices circulate, develop and interact in time and space, and about the forms of emergence, persistence and disappearance that follow. For the moment, and in the context of this book, the more immediate challenge is to articulate the relation between practices and changing patterns of demand for goods, resources and services. Whilst some theorists emphasise the 'immaterial' nature of (moral) practices (Taylor, 1989; MacIntyre, 2013), many draw attention to more 'material-economic' aspects (Kemmis, 2009). From this point of view, every social practice demands something whether that be a physically fit human being, an internet connection, a house, a flat surface or an electricity supply.

To give a simple example, walking in the park requires a body that can move, ground that can be walked upon and air that is breathable. It also depends upon, and contributes to, a long history of changing ideas about well-being, exercise, fresh air, nature and public provision. As well as needing leisure time, contemporary configurations of walking in the park suppose parallel infrastructures of path maintenance, watering and mowing. Other formulations are possible, and as the practice changes, so do the forms of demand and the systems of provision associated with it.

By implication, the material demands of practices are not fixed for all time: as practices evolve, some previously necessary elements become obsolete, some persist and new items are introduced. Trentmann, an historian of consumption, identifies long-term trends in the material composition of many practices, and argues it is this that underpins long-term trends in resource consumption (2016). For example, practices like washing clothes now involve the use of a machine and electrical rather than human power. Similarly, more equipment is needed, for instance for office working, parenting or cooking, especially when tasks become mechanised and when supporting infrastructures spread. In the energy sector and in transport

studies these ideas provide some clues as to how and why the resource intensity of everyday life is rising (Shove and Walker, 2014).

To go further and to understand the demand for particular resources, appliances or devices we need to address more specific questions about the roles things have in practice. We develop this topic in Chapter 3, but is already clear that there are different forms and temporalities of demand associated with 1) things that are used up in the course of conducting certain practices (resources such as food, electricity or petrol); 2) things that are directly engaged with in practice (devices, appliances, such as cookers, cars or light bulbs) and 3) things that are in the background (infrastructures, such as power grids or road networks) and that enable both the flow of resources and the functioning of devices. Not all practices depend on the conjunction of resources, devices and infrastructural arrangements but those that do have implications for demand in all three domains at once.

In thinking about the links between practices and the demands that follow it is important to recognise that it is often not the resource or consumer product that is, in a sense, demanded, but rather the services these material arrangements make possible. For example, heating a room depends on fuel, infrastructure and boiler, but it is their conjunction that counts in that it is the heated environment that enables different activities to take place indoors, and in comfort (see the discussion of services in Chapter 1).

Since daily life consists of many practices, and of complex connections and relations between them, understanding patterns of demand also depends on understanding how these interactions form and change. This is not entirely new territory. In transport studies, the concept of derived demand is, for instance, used to make the point that people travel not for its own sake, but in order to take part in activities separated in time and space (Mokhtarian and Salomon, 2001). This begs the further question of what it is that transport demand is derived from. Responses are often framed in terms of exogenous explanatory factors, such as 'the economy'. The problem is that such responses tell us very little about exactly what is going on. It is, for instance, true that levels of car ownership and the number of kilometres driven have increased with GDP (Bastian et al., 2016). But if we look more closely we can see that the precise nature of this relationship is not fixed: it changes over time. Similarly, if we go beyond the averages, it is clear that at any one point in time there are very different patterns of travel and of car dependence (DfT, 2018a).

Connections between specific social practices and observed demand (in this case miles or kilometres travelled) get lost in averaged and

aggregated data, but they are nonetheless there. As Mattioli and colleagues demonstrate, different methodologies can be used to identify and track the emergence of what they describe as 'car dependent' practices (Mattioli et al., 2016). As their research shows, practices like working, shopping and visiting friends and family do not always depend on the use of a car, but in some situations material arrangements such as city planning, road infrastructures and systems of public transport mean that cars and driving are indeed required (Shove et al., 2015). By implication, the resulting demand for fuel is not to be taken for granted, nor is it an expression of consumer choice, it is instead and in very specific ways, derived from how certain practices are configured and organised.

In short, demands for energy are provisionally held in place by a raft of social and technical arrangements, all of which have complicated and contested histories, all of which are open to negotiation and change and none of which are inevitable or natural. Nationally and internationally, these many end uses and practices combine to constitute the total demand for energy. However, this is still only part of the story. In thinking about how patterns of demand come to be as they are, it is clear that end uses are, to an extent, defined and made possible by the systems of provision and by the infrastructures, technologies and material arrangements on which they depend. This argues for a focus not only on consumers' practices, but also on related and connected forms and systems of provision.

Proposition 2: demand is made, not simply met

We are surely not the first to acknowledge the interweaving of supply and demand. In his seminal book, *Networks of Power*, Hughes (1993) describes the work that providers put in to building the demand for electricity, one practice at a time. In particular, new uses for electricity had to be established, and new appliances (cookers, fridges, hoovers, toasters) introduced to actively build demand at certain times of day and to manage otherwise very uneven load profiles. Taking a longer term view, Trentmann and others show how developments in domestic heating arrangements, like the provision of gas central heating, have co-evolved alongside ideas and practices of comfort. To be more precise, they have been part of establishing the concept of space heating, and of embedding expectations that homes should be heated or cooled to enable a uniform indoor temperature all year round (Trentmann and Carlsson-Hyslop, 2018; Rinkinen and Jalas, 2017). In this context, the convention of maintaining indoor climates at 18–22°C. is

arguably an outcome of the development and availability of mechanical heating and cooling systems capable of delivering these conditions (Cooper, 2002).

These are not localised or isolated processes. Instead, and as; Rinkinen et al. (2017) show, systems of provision and demand constitute each other across different scales and areas of daily life. In writing about frozen food systems these authors demonstrate that infrastructures (reliable electricity networks; systems of food distribution) and domestic appliances (refrigerators and fridge freezers) are jointly implicated in the emergence of increasingly global networks of food production and retailing, linked to also changing diets and methods of cooking, buying and storing food at home. As these examples suggest, demand is an outcome of relations between providers and institutions along with policy makers and consumers. In the chapters that follow we develop these ideas, and in so doing contribute to the work of those who conceptualise large social phenomena (Schatzki, 2016) as expressions and outcomes of social practices and relations between them.

Implications and challenges

This far we have suggested that social theories of practice provide the basis for a distinctive account of how demand is made and how it changes. Having introduced the first two of our propositions – that demand is derived from practices, and that demand is made, and not simply met – the rest of the book elaborates on the implications of this approach and the challenges and opportunities that it presents. Chapters 3, 4 and 5 unpack the practical and theoretical significance of our next three propositions: that demand is materially embedded; that demand is temporally unfolding and that demand is modified and modulated, deliberately or not, via many forms of policy and governance.

Before moving the discussion on, we bring this chapter to a close by reflecting on the power of already established approaches. If this book is to make a useful contribution and if it is to influence to climate change policies it will have to make its way in a field crowded with conflicting but influential ideas about behaviour change and the economic drivers of demand, and by traditions of policy making steeped in these paradigms. If we are to make any difference at all, we have to explain what current methods miss and what we have to offer. The simple answer to this is that our approach makes it possible and in fact necessary to engage with fundamental questions of

demand, and to do so 'head on'. By contrast, strategies that promote efficiency and behaviour change inadvertently reproduce interpretations of need and normality that are themselves part of the problem.

In the energy sector, policy makers and analysts routinely suppose that consumer behaviour can be nudged or changed through price signals, or by providing information that enables people to make better choices for themselves (Carmichael, 2019). Identical arguments underpin national and local government initiatives in transportation (e.g. persuading individuals to switch modes; modelling individual choice, etc.) (Marsden et al., 2014) and in other areas too. There is no shortage of academic literature that is broadly in support of such approaches, including research on the importance of identity politics, the need to foster green lifestyles (Hobson, 2003; Spaargaren, 2011) or the significance of consumption as a site of social and political expression (Holzer, 2006). Although this focus on behaviour and choice is entirely normal it is also limited.

Estimates vary but most researchers agree that measures designed to persuade people to change their ways make somewhere between 5 and 10% difference to the energy that they consume. This very limited impact makes perfect sense given what we know about how social practices are configured and organised, and about the forms of resource demand that follow. It is also consistent with the conclusion that despite being referred to as 'demand side' strategies, efforts to promote efficiency and behaviour change are not really intended to transform the range and character of services on which demand depends or to have a major or lasting impact on the fabric of daily life. Instead the emphasis is on trimming consumption and reducing waste whilst maintaining normal levels and standards of service.

This goes hand in hand with a much more pervasive reluctance to consider the extent and character of demand or to ask, up front, what is energy used for in society, how does this change, and how might it be steered? The UK's recent clean growth strategy (UK Government, 2017a) is not at all unusual in bypassing this topic entirely. Nor is it unusual in supposing that the central task is that of meeting present 'needs' with fewer carbon emissions than might otherwise be the case. Somewhat ironically, the politics of demand reduction entails a tacit commitment to preserving present standards of living, and to doing so at almost all costs. In this context, efforts to decarbonise supply and promote energy efficiency (that is to deliver the same or more service for less energy) are favoured because they promise to reduce consumption (or carbon emissions) but without challenging demand: that is without changing or modifying the range of practices enacted in

society today. In taking this approach, energy and climate change policies in particular take demand for granted, supposing either that it exists ready-made; that it is a non-negotiable 'need' and/or that it is made by uncontrollable market processes or influenced by policies in other fields (Marsden et al., 2014). This is not an especially favourable environment in which to make the case for a much more open and also much more challenging debate about the long-term sustainability of current expectations and standards.

One way forward is to point out that current policies, including so-called demand side management, or the promotion of energy efficiency, are not neutral: rather, they represent strong and powerful interventions in the constitution and perpetuation of demand and in the reproduction of resource intensive ways of life. By taking present standards as the benchmark, strategies to increase the energy efficiency of buildings preserve contemporary interpretations of comfort. Measures like these consequently have the important effect of stabilising and legitimising contemporary interpretations of normality and of excluding and obscuring opportunities for reducing demand at source.

At the same time, and since policy making is not of a piece, there are forms of intervention that inadvertently undermine energy and mobility intensive practices. For example, there are established and widely accepted methods of planning cities that limit the need to travel. Articulating the expectations and assumptions that are embedded in different sorts of policies and approaches represents a first step in really thinking about demand.

To go further depends on recognising the role of the state and of other actors in constituting demand and the means by which it is met. Amongst other things, this depends on noticing that the practices on which demand depends do not arise by chance: they have histories, and forms of interconnection, and they rely on material arrangements, infrastructures and systems that extend well beyond the individual consumer. This way of thinking challenges the very foundations of established discussions of ethical consumption and consumer choice, and in the same move establishes what one might think of as a politics of practice. For example, instead of worrying about what prompts individuals to drive to work, this analysis suggests that the more important policy challenge is that of understanding and influencing both the emergence and the disappearance of 'car-dependent' practices; or practices that depend on the use of a freezer; large amounts of data; high-speed broadband, or whatever.

In working towards this conclusion, which is also a starting point for the rest of the book, we have outlined different methods of

conceptualising demand. In the process we have shown how economic traditions and studies of consumer culture combine with policy discourses that obscure rather than reveal the underlying dynamics of demand. In response, we suggest that social theories of practice provide ways of conceptualising the interlinking of supply and demand as historical but also ongoing processes, and that they enable us to understand how multiple demands are constituted as practices change.

3 Matters of demand

This chapter explores the proposition that demands for energy and for services such as mobility are partly defined by and almost always linked to materials including resources (gas, water, oil); infrastructures (power grids, sewerage systems, road networks); and appliances, tools, devices and other objects (fridge freezers, cars, showers). In some senses, this is not a new or especially insightful observation. As one might expect, things figure prominently in many accounts of consumption and demand, and especially in discussions about the ecological consequences of consumer society.

Claims that society is becoming more and more materialistic are quite plainly related to the suggestion that there has been an influx of consumer products, that there is more 'matter' in circulation than ever before, and that peoples' identities are increasingly tied to forms of acquisition and ownership (Holt, 1995; Trentmann, 2016). This links to growing interest, especially in geography and economics, in the physical and environmental consequences of such trends. In these disciplines there is a strong tradition of tracking material flows and representing the global circulation of goods as they move through processes of production and consumption (Hardt et al., 2017).

As we explain in more detail below, efforts to describe and measure the metabolism of society usually work with stripped-down representations of flows, volumes and quantities of 'raw' materials. Although they represent the inputs and outputs of consumption and production, the history and the dynamics of demand and the social or cultural meanings of acquisition and use are usually missing. Instead, and, as in much economic theory, the 'need' for standardised product groups is treated as an outcome or expression of market forces. By contrast, and as outlined in Chapter 2, questions about how things acquire symbolic significance and value and how they figure in processes of trade and market exchange tend to be more prominent in fields like cultural

studies and anthropology. In much of this writing, and in part because of the focus on cultural variation, things are often treated as blank surfaces on which diverse and changing meanings and exchange values are inscribed (Appadurai, 1994).

Within science and technology studies and/or actor network theory there is an emphasis on things in action and in use. Rather than figuring as carriers of symbolic meaning or as standardised 'units' or traces of material flow and economic exchange, non-humans are conceptualised as active participants in social life (Latour, 1992). Various lines of enquiry follow. From the early 1990s onwards, actor-network theorists have described the constitutive part things play in making and shaping society (Latour, 2005; Verbeek, 2005; Wajcman, 2010). Within this field, much attention has been given to specific interfaces between humans and artefacts, showing how objects 'script' peoples' actions (Akrich, 1992); and how human-non-human hybrids are configured (classic references include: Woolgar, 1991; Haraway, 2006). This tradition goes further than recognising what Marx described as the 'use' value of objects and commodities in that at least part of the interest is in how human-non-human interactions form and change. Building on some of these insights, this chapter provides a distinctive and detailed account of how materials – ranging from specific artefacts through to large-scale technical systems – are implicated in establishing and transforming social practices, and hence in establishing and also transforming patterns of demand for energy and other resources.

Our first two propositions – 1) that demand is an outcome of social practice, and 2) that demand is made and not simply met – inform the third, 3) that demand is materially embedded. Elaborating on the suggestion that the demand for energy and mobility is integral to and inseparable from the array of practices of which social life is made, this chapter homes in on a series of related questions. In particular, how do material relations and arrangements combine in making and transforming social practices and the demands that follow? In general terms it is clear that things including road infrastructures, cars and petrol enable automobility and thus enable the demand for car-based travel (Urry, 2004). Similarly, the electricity network enables the demand for electrified home appliances, and for the power needed to make them work. In short, some material arrangements 'breed' demand in the sense that they rely upon or generate the need for related goods and services. Equally, infrastructures are of little or no value if there is no call for the resources and services they provide. In focusing on these interrelations we ask how different forms of demand – for infrastructures, for appliances and

for resources – develop together, and how all are jointly woven into what Schatzki (2011) describes as the 'plenum' of social practice: that is the sum total of all practices enacted in society.

In approaching these topics we distinguish between the different *roles* things play in the conduct of social practice. As we explain, some things (we call them resources) are used up or transformed in the course of doing certain practices: for instance, flour is used in making bread; petrol is consumed in driving a car. Others (we call them appliances) figure as instruments or artefacts that are interacted with directly, as people travel to work (car), cook dinner (oven) or relax (TV). Meanwhile, what we refer to as infrastructures are necessary arrangements that exist in the background: they are not interacted with directly, but they enable many practices at once. Examples include water systems; networks of power; road networks, the internet etc. In working with these ideas we adopt what Star (1999) describes as a *relational* approach: in other words we take the view that things do not have a *fixed* status, instead they acquire one or more of these roles depending on how they are positioned in relation to one or more practices. This conceptual move informs the ways in which we frame and address more complex questions about how interdependencies between multiple practices relate to, and arise from multiple forms of material interconnection.

In the second part of the chapter we introduce and illustrate the significance of this approach with reference to two empirical examples. The first concerns the emergence and changing status of a specific energy demanding appliance (the fridge freezer) as that is situated within also changing systems of food provisioning. The second example shows how travel demand is reconstituted through changing modes of shopping: some anchored in town centres and retail outlets, others depending on new configurations of internet, storage and distribution. Together, these cases give a sense of how material roles interweave and of the different forms of historical layering, codependence and prefiguration involved.

As is already obvious, this strategy differs on a number of counts from those that treat processes of production and consumption as phenomena in their own right, or that focus on non-human and human actors aside from the practices and processes in which both are enmeshed. In order to explain what is distinctive about our approach we briefly revisit what have become dominant methods of representing and describing the technologies and infrastructures involved in making and changing the need for energy and mobility.

Conceptualising energy and energy-demanding technologies: resources, artefacts and users

In physics, as in economics and environmental analysis, it is common to conceptualise resources as abstract commodities and to represent them in standardised terms. Different units are used to describe quantities of gas, electricity, oil or wood but in all cases the metrics involved enable forms of buying and selling on which energy and other markets depend (Callon et al., 2007). Similar techniques make it possible to follow the routes that diverse materials take as they travel from 'cradle to grave', and to evaluate the relative efficiencies of the different technologies encountered along the way. Sankey diagrams, such as in Figure 3.1, exemplify this approach. They show the paths taken by various energy sources (gas, coal, oil); the stages of conversion and transformation through which these pass, and the end services they enable.

Mapping flows in this way depends on the use of unambiguous methods of quantifying energy or of carbon emissions, without which it would be impossible to see how resources and commodities circulate through society, or to identify the proportions involved in providing heating, hygiene, mobility and so on. Agreed metrics are at the heart of disciplines like physics (Labanca, 2017) and economics, and are a precondition for many forms of environmental evaluation and comparison. For example, efforts to establish how close nations are to developing a circular economy require methods of calculating the 'quantity and quality of input flows, keeping track of interactions among system components across scales, and identifying environmental costs and savings of loop-closing strategies at all levels' (Geng et al., 2013). Other commentators focus not on nation states but on specific regions or cities. For example, those who write about urban metabolism conceptualise 'the city' as a compilation of different sociotechnical systems, which 'source, use, and transform huge amounts of natural resources' (Monstadt, 2009). Developing this idea, Monstadt argues that cities constitute 'pivotal sites at which the resource flows "metabolized" by infrastructures are geographically concentrated, and the everyday exchanges between networked infrastructures and natural environments occur' (Monstadt, 2009). In this analysis, megacities act as 'nodal points of global economic, cultural and political networks' (Castells, 1996) through which resources are organised and channelled. Building on these ideas, other commentators follow the lives, and life cycles, of specific objects, comparing processes of production, distribution and appropriation and recognising that environmental impact is scattered across all these stages (Tukker, 2000; Saccani et al., 2017).

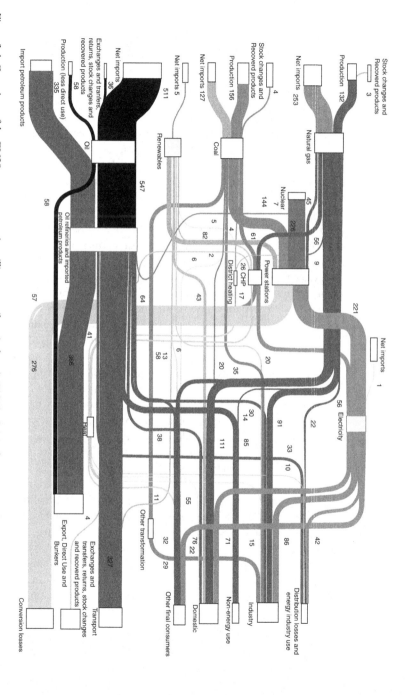

Figure 3.1 Overview of the EU28 energy system in million tons oil equivalent (MTOE). Original source: European Environment Agency (EEA).

All such representations depend on forms of abstraction, that is, they depend on stripping resources out of context, and obscuring the distinctive social and historical processes in which material relations and shifting patterns of consumption are enmeshed. Whatever the scale of analysis, be that nation states, a specific city, or an individual product, the emphasis is on which resources are consumed, to what extent, where and with what consequence. As a result there is little or no reference to the history of these arrangements (Shove, 2017), or to the changing constitution of the 'services' that these flows enable. As in the Sankey diagram in Figure 3.1, categories and concepts of service are taken to be fixed: what matters, and what varies, is how these needs are met and the mix of fuels involved.

These strategies are symptomatic of a much broader tendency which is to suppose that technologies and the needs they enable are analytically separable. This conceptual move underpins many methods and lines of enquiry. For example, energy analysts routinely begin by describing different energy sources (coal, wind, bio energy, etc.); then considering technologies of generation and distribution (smart grids, engines, etc.), and the appliances involved (heating systems, powered devices), but largely overlooking the ways in which these arrangements co-constitute interpretations of need, including meanings of comfort, hygiene, convenience etc. (IEA, 2018; WEC, 2018).

As any more historical approach would show, social practices, energy systems and material arrangements develop together. In skating over these interdependencies and the forms of mutual influence at stake, dominant resource-based discourses are routinely cut short. More specifically, they fail to grasp 'why people use resources, how these "needs" and "wants" are constituted and how they are changing within the broader context of everyday life' (Strengers, 2011: 36).

The focus on resources, in the abstract, has other consequences. One is to favour what are often simplified accounts of how patterns of consumption change and of the potential for technological substitution. From a resource-centric point of view, carbon emissions can be reduced by swapping coal-fired technologies for those that depend on renewable energy; or by switching to more efficient methods of energy conversion. But as Rinkinen (2018) and Kuijer and Watson (2017) show, substitutions are rarely that simple. Historical research by these authors demonstrates that the move from coal- or wood-based heating to gas had many far-reaching implications for how households organise their day, for which parts of the home are used and when, and for the physical labour and the systems of provision involved (Kuijer and De Jong, 2012; Kuijer and Watson, 2017; Rinkinen,

2015). Put differently, material arrangements, whether these be gas stoves or log piles, have an active role in configuring what people do, not only in terms of heating, but also in the timing and location of many other practices as well.

These links and connections are not visible in studies that work with standardised units of energy or of carbon, or that focus on forms of technological innovation, narrowly defined. However, they are critical for understanding resource flows as outcomes of the processes through which things come to have meaning and acquire significance not as phenomena in their own right, but as part of what the anthropologist Tim Ingold calls 'form-making processes' (2012). In an account of what he refers to as an 'ecology of materials' Ingold advises researchers to turn from the 'objectness' of things to the task of accounting for the formative processes through which they come into being. As he explains, in a world of materials, nothing is ever finished: 'everything may be something, but being something is always on the way to becoming something else' (Ingold, 2011: 3). By implication, the same 'thing' can figure in very different ways depending on where and how it fits within different practices. More than that, understanding these always changing positions becomes an essential part of understanding the configuration and also the circulation of fuels and other consumer goods.

If taken to heart, these relational views of the material world complicate established methods of resource accounting. At the same time, they make it possible to link the flow of energy and resources through society to the making of demand, and to show how demand and materiality interact in practice. From this starting point, other questions arise. In particular, how do materials of all types (resources, appliances, infrastructures) figure in the emergence and transformation of different, but often linked complexes of social practice?

In everyday life, the consumption of gas or electricity is almost always mediated by infrastructures (wires, pipes, etc.) and appliances. The need for resources consequently depends on the range of energy-dependent devices in existence and in use, and on when and how these are mobilised in daily life. From this perspective, human-appliance interfaces are of special interest for an analysis of demand not simply because these are points at which energy is evidently 'used' but because appliances also help constitute what people do.

This observation is consistent with the contention that things are embedded in social relations: as Latour and Akrich explain, they 'script' (Akrich, 1992) and structure everyday activity; they enable forms of delegation and 'action at a distance' and they embody social

and political relations, past and present (Latour, 1992: 169). These core claims lay the ground for important lines of enquiry, for instance about how powered artefacts take the place of human labour (actions are delegated to energy-consuming devices); how infrastructures enable new connections between people and organisations (frozen food chains link distant consumers and providers); and how they relate to coexisting sociotechnical configurations.

As one might expect, there are different ways of thinking about the character of human and non-human interaction. At one extreme, proponents of what are known as posthumanist positions contend that humans are technological in the sense that they have fused together with machines over time (Haraway, 2006). Others retain concepts like those of 'technology', 'artefact' and 'user' and seek to describe how these analytically separate entities interact (Schot et al., 2016). Either way, there is a tendency to examine connections between people/humans/users and things rather than to provide a more encompassing account of the social practices in which both have a constitutive part to play (Schatzki, 2002). In other words, the challenge of understanding how material relations shape the very practices in which people are engaged, and how such practices emerge and extend over space and time is typically beyond the scope of 'user' studies however broadly defined. To address these issues we need to change tack and explore other ways of thinking about how materials of all kinds figure in the conduct and also the dynamics of social practice.

Reconceptualising energy and energy-demanding technologies: material relations in practice

Our first proposition was that demand is an outcome of social practice and our second was that demand is made and not simply met. We now contend that practices are embedded within, and inextricable from multiple material relations, hence proposition 3: demand is materially embedded. To develop a more precise account of what this means we need to say more about the material aspects of social practice. In the *Dynamics of Social Practice;* Shove et al. (2012) take materials to be elements of practices, along with meanings and forms of competence. This is a plausible strategy but it is also very unspecific: the category of 'materials' can encompass pretty much anything from the human body and the air that people breathe through to the pencils with which they draw, or the planes in which they travel. Schatzki, who also writes about the status of materials in relation to social practices

works with a different conceptualisation of material arrangements, but one that is also very broad. In his words:

> Human coexistence is inherently tied, not just to practices but also to material arrangements. Indeed, social life, as indicated, always transpires as part of a *mesh* of practices and arrangements: practices are carried on amid and determinative of, while also dependent on and altered by, material arrangements.
>
> (Schatzki, 2010: 130, emphasis in original)

Whether we see materials as integral to the conduct of practice, or as arrangements within which practices transpire, there is clearly scope for developing a more differentiated analysis of the material world, especially as that matters for patterns of demand. One way forward is to consider the *relation* between things and the practices they enable or of which they are a part. This method draws on some of Ingold's ideas and on Star's (1999) view of infrastructure, already mentioned above. In an article entitled 'The Ethnography of Infrastructure' Star claims that 'infrastructure is a fundamentally relational concept' and that things only become 'real infrastructure in relation to organized practices' (Star, 1999; 380). Whether something figures as infrastructure, or for that matter as an appliance or resource, consequently depends on how it is situated in relation to one or more practices.

Putting these pieces together, and doing so with issues of demand in mind, we now elaborate on three genres of material relation. The first includes things that are not interacted with directly but that are nonetheless required in the sense that multiple practices depend on them, and on their existence 'in the background'. What are classically thought of as infrastructures, including networks through which resources, goods and sometimes also people and forms of competence flow (Hughes, 1993) are frequently designed to function 'behind the scenes', to accommodate peaks in demand and to ensure reliable supply and uninterrupted service. Despite appearances, such arrangements do not simply meet existing needs: they also shape the practices they make possible. In very practical terms, systems that act as infrastructures are designed, sized, operated and repaired with certain expectations of demand in mind. In addition, they are inevitably constrained and unequal in the opportunities they afford (McFarlane, 2011; Graham and Marvin, 2001). Both features remind us that infrastructures are not simply inert arrangements of concrete, wires, tarmac or pipework, they are 'living' systems the qualities and potential of

which ebb and flow depending, amongst other things, on the geographies and timings of demand.

The normally invisible range of activities that infrastructures enable becomes apparent when power systems fail, and when it is impossible to use the many mediating appliances and devices via which resources and practices are linked (Nye, 2010). More than that, when the power is off, interconnections between things as varied as mobile phones, water supplies, gas central heating, traffic lights and cash machines become apparent. As is obvious in these moments, and in general, infrastructures make sense and have value not in their own right but because they provide the resources required by a range of different appliances, each of which is, in turn, linked to the conduct of one or more social practices.

Whereas infrastructures typically support and are part of several practices at once, appliances and devices frequently have a narrower role. This is not a clear-cut distinction: devices are often appropriated in unexpected ways. For example, although it is designed for drying hair, a hair dryer can also be used to help glue set faster, or to soften plastic. Equally, some devices are decidedly multipurpose. For instance, a smart phone enables numerous activities, not just phoning but also texting, streaming, taking pictures and playing games. The difference is that whilst infrastructures exist in the background and are not interacted with directly, devices and appliances are literally mobilised in practice. As such appliances and infrastructures stand in different relation to resources, here defined as things that are used up, or transformed in the course of practice. In general infrastructures enable the circulation or delivery of resources whereas appliances involve their use or transformation: for instance, an electric shower uses up or more accurately transforms electricity and water.

In combination, these observations suggest that understanding resource demand is, in part, a matter of understanding what energy or mobility are 'for'. More specifically, to which practices or complexes of practice do specific configurations of resources, appliances and infrastructures relate, and of which are they a part? To continue with the example of showering, having a bathroom equipped with a shower is neither inevitable, accidental or natural. Instead, and as the history of showering demonstrates the availability of running water, and the means to make it hot both precede showering as a practice (Hand et al., 2005). Equally, the shower and its related infrastructures do not, of their own, constitute showering as a daily routine: concepts of freshness and well-being are also needed. But when showering does become a habit, material roles and relations between resources,

appliances, infrastructures fall into a recognisable pattern, evident in the demand for water and for power and in when this occurs.

These recursive relations exist at many scales. Electricity infrastructures are typically designed and sized to enable certain levels of peak demand (Hughes, 1993; Shove, 2018). But if such systems are to operate efficiently, it makes sense to develop or to find ways of using the power that is available at off-peak times of day. In the early years of the electricity network in the UK, home appliances including hoovers and toasters were deliberately designed and introduced with this ambition in mind (Forty, 1986). In this case, a glut of resources (electricity) linked to features of infrastructural design (catering for peak load) prompted the introduction of new appliances in the hope that new energy demanding practices would follow, resulting in a more efficient load profile overall. The detail of how such configurations emerge and change varies case by case. Thus the material relations that constitute central heating, and the details of how these come together are not the same as those that define hoovering or mowing the lawn. Nor are any of these connections fixed for all time: material roles of the type we have described are interconnected, relative and overlapping.

In the next two sections, we develop and work with an account of material relations in practice that allows us to show how demands for energy and for mobility are established and how they change in two specific situations. Each case illuminates different aspects of our approach. The first shows how food and electricity infrastructures and systems of provision combine via patterns of eating and cooking enabled by the fridge freezer. The second examines the material relations of shopping, highlighting the multiple interlocking trends involved and their consequences for the mobility of goods and people. Both examples show how material configurations, practices and emergent patterns of demand develop and change together.

Example 1: Fridges, frozen food and electricity – interdependent demands

It is no wonder that fridges and fridge freezers account for a significant proportion of domestic energy consumption. These appliances are now owned by all but a very few households in the Global North, and by many in the Global South as well, and they are switched on all the time. This has not always been the case, nor is it a natural or even inevitable state of affairs.

Prior to the 1940s, home refrigerators were relatively rare even in the USA. So what has changed? What accounts for the relatively rapid uptake of these devices, for how are they embedded in domestic and commercial practices and for how this status is established and maintained? Fridges keep food cool, but freezing, which preserves food for much longer, has a different role in the nexus of provisioning practices. More specifically, freezers enable consumers to become part of a global food system in which goods travel very long distances, in which they are supplied and consumed 'out of season', and in which forms of preparation and labour are outsourced, as is the case with ready meals. The widespread need for fridges and especially for freezers is arguably linked to the geographical separation of agricultural production and food consumption, and to the status of certain types of food. Dupuis' classic history of milk consumption shows that the political, commercial and cultural positioning of what became known as 'nature's perfect food' depended, in no small measure, on the potential for mechanical cooling (Dupuis, 2002; de La Bruhèze and van Otterloo, 2004). Today, what is known as the cold chain permits the worldwide circulation, storage and transportation of a very wide range of meat and dairy products, and of other foodstuffs as well.

We argue that fridges and freezers are now as dependent on these extended systems of food provisioning as they are on a consistent and reliable supply of electrical power. More than that, we claim that understanding how domestic energy demand has increased depends, in this case, on understanding how refrigeration and freezing fits within, and also shapes a landscape of changing practices relating to eating, cooking and storing food.

From this point of view, tracking the number of appliances in use, or focusing on their energy efficiency ratings tells us very little about how demand for cooling is constituted and reproduced. Instead, what is required is an analysis of how conjunctions of diet and practices of cooking and provisioning have cohered around fridges and freezers. This is important in that not all such conjunctions are the same. In writing about fridges and freezers in the UK, Shove and Southerton (2000) describe three stylised moments in freezing history: one in which freezers were for storing gluts of home produce; one in which they enabled bulk buying and one in which they figured as time-saving appliances. In each of these moments, households, freezers and food providers are linked in different ways with implications for what freezers actually contain, and for their status as essential or optional appliances within the home. These configurations are, in turn coloured by changing discourses and ideologies of care, convenience,

health, well-being and family life (Hand and Shove, 2007; Shove and Southerton, 2000).

As these sociological studies of freezing show, specific combinations of commercial food provision (selling ready meals, offering frozen food in supermarkets) appear to have created the need for a freezer in homes that had not needed one before. However, this is not a one-way story. Domestic freezers represent an essential last link without which the rest of the commercial cold chain makes little or no sense. From this point of view, home freezers are best understood as *part of* the cold infrastructure on which so much food provisioning now depends. In other words fridge freezers simultaneously figure as home appliances (Shove and Pantzar, 2005; Shove et al., 2012), and as essential infrastructure that has enabled changing systems of food production and provision (Freidberg, 2015). More than that, the freezer's device-oriented role within the home defines and depends on its infrastructural roles within practices of production and distribution. Acknowledging this duality enables us to detect the interpenetration of material relations that connect and also transform practices of cooking and eating along with those of growing, manufacturing and distributing food.

This account suggests that around 20% of UK domestic electricity demand is directly tied to the ways in which food is bought, sold and stored and to what foods are consumed and cooked. This takes different and always evolving forms. Staying with the cold chain, a recent study in Hanoi and Bangkok found that whilst some households relied on fridges and freezers to keep and store what they describe as Western foods others depended on them to help maintain a much more traditional diet (Rinkinen et al., 2017). There is no one fixed model, but in all cases resource consumption not only in the home, but also in the provision of refrigerated transport, cold stores, manufacturing plant and production facilities around the world, is an outcome and an expression of specific configurations of multiple infrastructures and appliances as these are mobilised in practice.

The next case suggests that the changing demand for mobility, both of people and of goods, can be analysed and understood in similar terms.

Example 2: On- and off-line shopping – reconfiguring material relations, practices and travel demands

Figuring out how demands take hold is a matter of explaining how combinations of practices and material relations interlink and change

'en masse' (Shove et al., 2015). As the previous example suggests, developments in energy demand are rarely the result of some isolated innovation: instead, demands emerge and change through the conjunction of domestic and commercial practices, sometimes involving multiple infrastructures as well. This is an important insight in that it forces us to engage with fundamental questions about how social and material relations combine at scale.

As our second example shows, the organisation of in-store and online shopping, the configuration of these forms, and the changing relation between them has practical consequences for movement of people and goods. What shopping involves, and when, how and where it is done is in turn defined by a multitude of historical and contemporary arrangements, including aspects of urban design, systems of automobility and the organisation of commercial as well as 'consumer' logistics (Cochoy et al., 2015). Different modes of shopping (which encompass the practices of selecting and purchasing things, and sometimes bringing them home) develop at the intersection of systems of provision and consumption and also link these systems together (Stobart, 2010).

In this context, the rise of what is known as 'online' shopping reflects and depends upon novel combinations of practices and material relations. Rather than going to the shops themselves, consumers view and buy goods at home or on the move and have them delivered to the door. On the supply side, online shopping has led to new organisational forms, including links between providers, retailers and other actors as supply chains and delivery systems are rearranged. Out of these conjunctions, new patterns of movement emerge, both for consumers and for those involved in producing and selling the things they buy.

In the UK, online shopping now represents almost 17% of total retail sales. This has coincided with a 30% decrease in physical shopping trips over the past two decades but a 25% increase in distance travelled by vans (Allen et al., 2017). It is tempting to interpret these figures as evidence of the impact online shopping has had on travel demand. But rather than attributing shifting patterns of mobility (of people and of goods) to an apparently single cause (online shopping) we treat them as consequences of multiple *interlocking* trends in how, where and when goods are manufactured, bought, sold and used.

Taking a slightly longer-term view, it is obvious that there is no one dominant or unchanging form of shopping. For example, as supermarkets and big retail outlets became more dominant in the UK, the car dependence of shopping and the distances travelled rose overall

(Mattioli et al., 2016). Equally, the rise of out-of-town shopping was itself made possible by systems of automobility, linked to philosophies of urban development and economic growth.

Analytically, out-of-town retailing and online shopping are alike in that both are situated at the intersection of multiple infrastructures: variously including provision for the car, town planning and more recently internet access and broadband networks. Neither can be attributed to just one innovation (e.g. apps for online shopping, the size of the car boot; the shopping trolley): instead, both are woven into a more complex texture of continuity, hybridisation and sometimes radical reconfiguration in which linked systems and infrastructures develop together, but at different rates and over different periods of time.

It is evident that there are coexisting configurations of online shopping, aspects of which are shared with in-store variants. Whilst some material arrangements are definitely new, and some are also unique to buying and selling online, there is extensive interaction for instance, between automated warehouses, remote locker boxes and established infrastructures that are being used in different ways (stores offering click and collect, shops repurposing stock rooms, etc). So although online shopping is growing rapidly, it is not a simple question of being either online or in-store. Practices of shopping are instead extended to encompass browsing, comparing, purchasing and receiving goods, each of which can happen in different ways and in different locations. In all of this, the materiality of what it is that is being bought remains important. There are significant differences in how various aspects of shopping work out in relation to highly standardised or customised goods, to those that are repeat or one-off purchases, that have some aesthetic quality or that are exceptionally heavy.

Thinking about shopping not as a bounded phenomenon but as a complex of social and material arrangements allows us to consider the distinctive spatial and temporal features of different modes and the extent and character of the energy demands that follow. It also allows us to recognise that certain infrastructural forms, such as 'high streets', persist for much longer than the complexes of practices of which they were once a part. People are consequently surrounded by the remains of previously networked material arrangements that are no longer sustained by the bonds of interdependence that used to hold them in place. As a result, forms of disappearance and emergence are often partial in that contemporary complexes of practice frequently depend on modified or existing material configurations, blended in new combinations.

These brief discussions represent worked examples of how specific forms of demand – in the first case for energy (to keep food cool) and in the second for mobility (linked to different configurations of shopping) – arise and change. These instances are not representative of all possible forms of demand-making, but the points they raise and illustrate apply to other situations as well. In particular, these narratives underscore the importance of recognising that infrastructures, systems of provision, divisions of responsibility and complexes of practice change *together*. Second, and because of these interconnections, what seem to be solid arrangements of supply and demand are inherently dynamic: previous configurations are not the same as those that exist today, and future combinations are likely to be different again. Third, and critically, the extent and character of demand, whether for energy or mobility, is an *outcome* of these extensive, linked and shifting arrangements. In the next and final section of the chapter we reflect on the wider implications of these conclusions.

Implications for demand: persistence, planning and prefiguration

This far, it might seem as if material relations, practices and patterns of demand simply emerge. This is not always the case. Often, what happens is in part related to variously deliberate forms of planning, adaptation or anticipation. Not all interventions have the desired effect, but it is important to take note of the fact that the systems and arrangements we describe do not arise by chance alone.

Sometimes policy makers and organisations focus on *adapting* established infrastructures to suit new roles and purposes. To some extent this is what is happening on the high street, as retailers and planners accommodate to the realities of online shopping. The rise and subsequent decline in the number of filling stations represents another case of adjustment, in this case linked to the efficiency and reliability of the vehicle fleet. In the UK in 1970 there were 20,000 filling stations and around 12 million cars (Mackay et al., 2003). By 2018, the number of cars had increased to 30 million yet the number of filling stations had declined to 8,500 (UKPIA, 2019). Many are now reconfiguring as mini supermarkets, coffee shops and fast-food chains. As this example suggests, existing infrastructures are sometimes incorporated into the provision of new or expanded services. Similar forms of repurposing have occurred as old military airstrips have been brought back into use by Ryanair (a low cost airline) in response to and as part of making demand for air travel. The reinvention of the canal

network as a leisure facility rather than a freight transport system is a similar example. In each case, new complexes of practice form around the remains of material arrangements that have been developed, and in some cases previously rearranged, through a succession of what are often piecemeal reconfigurations.

Although they are rarely established entirely from scratch, certain forms of infrastructural investment are more obviously designed to meet what are taken to be present or future needs. This is so in transport, in road building and in the construction of new power stations, and in these contexts there is an established tradition of what is known as 'predict and provide' (Goulden et al., 2014). As the term suggests, policy makers and providers aim to design systems of provision and supply that are capable of meeting anticipated demand. In Chapter 5 we discuss some of the methods used to forecast future need, but in the present context, what matters is the fact that techniques of this kind are 'performative': in other words they are themselves instrumental in constructing and enabling future increases in demand. In very practical terms, the logic of predict and provide results in infrastructures that are sized to cope with expected peaks in load – making it possible to 'keep the lights on' at all times, or to keep the 'wheels' of industry turning, come what may.

This is important in that actual and also planned infrastructures consequently prefigure social practices and relations between them, meaning that they have a forward looking aspect, shaping, influencing and affecting the social future (Schatzki, 2010). The same applies to plans and programmes of repair. Keeping infrastructures going takes work, and in putting that work in, the organisations involved reproduce and enable expectations not only of present but also of future demand.

Although histories of infrastructures – electricity networks, road infrastructures, information networks – are almost always stories about the practicalities and politics of provision and supply, this is only part of the picture. As represented here, there are other tales to tell about how infrastructures enable shifts in practices, and in relations between practices, and thus in the patterns of demand that follow. It is already clear that this is a patchy and complex process and that the mere existence of appropriate material arrangements does not, of itself, mean that expected patterns of consumption will ensue. Many practices suppose the coexistence of multiple infrastructures, not one alone, and for this and other reasons, the trajectories of social practices are not equivalent to the trajectories of sociotechnical arrangements: an observation that complicates otherwise plausible accounts of technological path dependence (David, 1985), and of 'transition' theories (Geels and Schot, 2007).

At the same time, there is no denying the point that massive increases in CO_2 emissions over the last few decades are outcomes of what Schatzki describes as 'practice-arrangement' nexuses (Shove, 2017), and that current ways of living involve unsustainable configurations of resources, infrastructures and appliances. In this chapter we have added to the repertoire of concepts that can be mobilised to understand trends at this scale. Focusing on how material relations and practices develop together, and over time, takes us beyond the normal discourses of technological efficiency, consumer choice and price. More than that, our account provides a means of thinking about material flows, but without abstracting resources as if they were significant and worth following in their own right. One problem is that this far we have discussed material relations, practices and demand, without any reference to matters of time and timing, or to when resources are consumed. Since this is an increasingly important issue both in the energy sector, and in relation to transport, it is the central topic of Chapter 4.

4   The temporalities of demand

Demand, whether for water, mobility, electricity or gas, is typically represented in the form of total or average annual consumption, measured in standardised units like litres of water, cubic metres of gas, kWh or passenger kilometres travelled. Although figures of this kind give a sense of extent and scale, they obscure temporal variations in the relationship between supply and demand. In focusing on these fluctuations and their significance for the design, management and operation of systems of provision, and for attempts to modify demand, this chapter shows how such relations are embedded within and also constitutive of the social and temporal organisation of daily life.

There are various reasons why the temporal aspect is important. One is that networks and systems are generally designed and sized to cope with peaks in demand, meaning that there is excess capacity at other times. Decisions about precisely how much supply is required and how much of the peak could or should be met are always complicated. In the electricity sector, providing infrastructure that lies idle for 75% of the time increases costs across the system as a whole. On the other hand, the philosophy of 'predict and provide' reflects a political as well as a commercial commitment to maintaining supply, and to 'keeping the lights on' at almost any price. The resulting tensions are discussed by Strengers et al. who observe that in Australia:

> A quarter of the entire electricity network is only needed for a few hours on very hot days so that people can run their air conditioners (on top of other demand). This rarely used infrastructure is being paid for by everyone.
>
> (Strengers et al., 2014: unnumbered)

A somewhat different approach exists in the transport network where congestion on roads and overcrowding on public transport systems reflects the impossibility of swiftly altering supply to cater for demand, despite political rhetoric surrounding supply side solutions (Goulden et al., 2014). In this context, there are tacit and sometimes explicit understandings of acceptable levels of congestion, and of thresholds beyond which interpretations of normal service are compromised. Maintaining provision within these limits represents a goal much like that of keeping the lights on and, as with the electricity network, sizing for the peak also results in substantial overcapacity for the majority of the day and in particular at night. This in turn creates commercial incentives for service innovations to stimulate demand during off-peak periods.

In transport as in the energy sector there is a tendency to anticipate growth and add capacity to accommodate future demand, but this is not the only option. What are known as methods of 'demand side management' are designed to shave present and future peaks and thus reduce or avoid the need for extra, more expensive, or more carbon intensive supply. Whichever path is taken, questions about exactly how much capacity to provide, and about when, whether and how demand might be deliberately scaled back bring more fundamental issues into view about the timing of supply and demand, and about how peaks and troughs arise in the first place.

A second reason why issues of time and timing are rising up the agenda is that the energy supply mix is changing. Unlike conventional fossil fuels that are available all year round and that can be stockpiled and stored, wind and solar power are much more intermittent. The challenge of decarbonising electricity systems and increasing reliance on renewables means that it is ever more important to know not only about the scale of demand, but also about exactly when power is used. If we take a long view, the problem of matching demand to intermittent supply is not a new concern. As Kris de Decker explains, in the past, millers would only work, and corn would only be ground on windy days (De Decker, 2009). Looking ahead, we would do well to ask whether future demand might be similarly timed and tailored to fit more fluctuating and perhaps more unpredictable patterns of renewable energy supply. What are the implications of such an approach compared with alternatives like investing in storage to overcome temporal fluctuations, and where should the balance between these options lie?

Third, and more abstractly, a discussion of the interlocking temporalities of supply and demand forces us to consider the societal

scheduling of social practices. It is obvious that load profiles including those of electricity demand or of congestion on the roads mirror the social and temporal organisation of everyday life but how do these arrangements come about? (Shove, 2009). As will become clear, these questions call for further analysis and discussion of the duration and sequencing of what people do, of the forms of societal synchronisation involved, and of how temporal patterns emerge and change within and between societies and over generations and decades.

This is a matter of practical as well as theoretical significance. In their day-to-day work, people responsible for managing utilities, private companies and public sector institutions confront the ongoing challenges of coordinating and scheduling deliveries, ensuring that resources are available and handling daily and seasonal variations in real time. Zerubavel's account of time patterns in hospital life (Zerubavel, 1979) gives a sense of the organisational work involved in administering medicines 'on time', and in combining the temporal rhythms of staff with those of patients. In other sectors, techniques like 'just in time' delivery are crucial for car manufacturing and for the supply of fresh fruit and vegetables. In all these situations, real time equates to clock time, defined in hours, minutes, days or months. This makes sense, but if we are to understand what lies behind these arrangements, and how patterns of demand come to be bunched or spread, we need to think about time in a different way.

There is an extensive sociological literature on time, ranging from the history of timekeeping (Glennie and Thrift, 2009) through to more philosophical questions about the relation between clock time and time as experienced, and as a dynamic and also emergent concept (Adam, 1990; Lefebvre, 2004). Rather than seeing time as a measurable and finite resource, some argue that time is, in a sense, made by social practice. This might sound odd, but it is evident that social and institutional rhythms form and change in ways that are closely related to the duration, the periodicity, the sequencing and the synchronisation of different activities (Zerubavel, 1979; Blue, 2017). To give a very simple example, the 'rush hour' is not an hour like any other. It is instead an hour, or more likely several hours, defined and in a sense made by the conventional timing and duration of the working day.

This argues for a less objective and a more relational view of time, along the lines of Southerton's representation of domestic time management. In writing about the scheduling of family life, Southerton introduces the notion that people 'squeeze' and 'stretch' time (Southerton, 2003). He uses these terms to describe how households pack

many practices into certain times of day in order to create other slower, calmer periods of so-called quality time. Similar strategies are reported by Thompson who writes about domestic schedules as outcomes of the ongoing juggling of competing priorities (Thompson, 1996). Kaufmann (1998) also draws attention to senses of obligation that generate what are experienced as compulsions to act at a particular moment. The examples Kaufmann gives include 'injunctions' such as the need to do the laundry before the pile of dirty linen gets too high or to wash the dishes after every meal. In combination, observations like these demonstrate that opportunities for squeezing or stretching time and for modifying interlocking forms of temporal organisation are not simply or strictly tied to clocks and calendars. This is important in that the scope for reducing periods of peak demand depends on the potential for rescheduling energy demanding activities and for reconfiguring the forms of sequencing and synchronisation of which socio-temporal rhythms are made.

By implication, and somewhat ironically, conceptualising the fluctuating relation between supply and demand (in clock time) calls for an understanding of time not as a fixed resource but as an outcome and a feature of the 'the creative production, reproduction and consumption of multiple temporalities' (Shove et al., 2009b: 3). In other words, it is what people do at different times of day and year that gives time frames like the week or the weekend meaning and that underpins experiences of being rushed, or pushed for time (Zerubavel, 1985). The rest of this chapter works with these ideas to provide a distinctive account of how and when peaks in demand arise and of the forms of flexibility that follow. The next three sections approach these questions from different angles.

Building on the previous chapter, we start with a discussion of the various temporalities associated with the provision and appropriation of resources, infrastructures and devices. Focusing on these material relations allows us to distinguish between resources that can be stored (or not); between networked and more intermittent forms of service provision and between large- and small-scale systems, all of which are relevant for the timing and duration of specific practices, and so also for the timing of supply and demand. Other considerations come into play in understanding how practices combine to form peaks and troughs in demand. In addressing these topics we focus on forms of societal synchronisation, and on how the many practices that make up a day, a week or a year are sequenced and scheduled, not in isolation but always in relation to each other. Our final move is to revisit the concept of flexibility. In the electricity sector, flexibility is variously

treated as a commodity, a resource that can be turned up or down, or as a characteristic of an energy system, a person or a specific practice. By contrast, we contend that the timing of supply and demand and the scope for modifying either depends on the interlinking of multiple social practices at a societal scale. These strategies allow us to elaborate on the proposition that demand is temporally unfolding and to set the scene for Chapter 5, and a more detailed discussion of the limits and possibilities of intervention.

Material relations, practices and the timing of demand

As a first step, we take note of the temporal implications of the many material relations on which energy and mobility demand depends. Some of these have to do with the perishability of different resources, and the potential to store them. In the food sector, 'use by' dates remind us that many of the goods we consume have a limited shelf life: by contrast, resources such as coal or oil can be kept for centuries on end. Infrastructures, including road or rail networks, have different temporalities inscribed in them, being constructed to last for so many years providing they receive periodic maintenance and repair. Meanwhile, domestic appliances like fridges or washing machines are designed to be always on; used intermittently or run for a certain number of hours per year. Other relevant features include the aspects of duration (how long does it take to run a short wash cycle?) and capacity (how many place settings does the dishwasher take?) (Wilhite and Lutzenhiser, 1999; Morley and Shove, 2014). In short, homes are designed and equipped in ways that embody judgements about both the volume and the timing of demand.

Similar considerations apply at an infrastructural scale. Although it is obvious that issues of weight, volume, ownership and price influence the extent to which things like petrol, gas or coal are stored and kept, these arrangements are not determined by material considerations alone. In addition, and in thinking about how gas or electricity networks handle and buffer fluctuations in demand it is important to recognise that these systems are not inert. Whilst it is relevant to know about the number of kV that power transmission lines can carry, figures like these do not reveal much about the capacity of the network as a whole. This is partly because the detail of how and when resources flow through infrastructures depends on how systems are operated and organised, and on the roles and responsibilities of those involved.

Histories of ownership combined with political commitments to state or private provision are crucial for interpretations of capacity and for precisely how fluctuations in supply and demand are handled in real time (Graham and Marvin, 2001). For example, in one of the electricity network control rooms that Silvast describes, operators are involved in dispatching loads, switching from higher to lower voltages, shutting down parts of the grid for repair, and still 'keeping the lights on' at all times (Silvast, 2018). In the room next door, different but related kinds of network management go on, as staff bid for power from the ever-changing Nordic pool (the energy market) in order to meet changing levels of demand. Sometimes breakdowns occur and power supplies fail but more commonly the challenges are those of making sure that supply systems accommodate peaks in demand, as and when they arise.

Similar strategies are adopted in the gas sector. At any one point, the network of gas pipes and their size is fixed. However, the amount of gas delivered, and when, depends on the careful and creative, management of pressure within the network. As Forman explains, network operators pack pipes with gas in anticipation of cold snaps and extra demand (Forman, 2017). This means that with gas, as with electricity, capacity proves to be a dynamic concept: an outcome of juggling and management, mediated by variation within the customer base (if this is large, differences in the timing of resource demand cancel each other out across the population), as well as by the material qualities of the infrastructure itself.

The effective capacity of road systems is also fluid. As we know, it is not just a matter of road widths and routes: traffic speeds drop when more people are on the move, and commuting practices adapt as the material affordances, not the material properties, of systems evolve. Figuring out how much strain the road network can take is consequently complicated. It is so in that answers depend on the level of service, including the speed of travel or the number of hours spent commuting deemed to be acceptable. Goodwin and Lyons (2009) show that interpretations of service vary over time and between cohorts, and depend on the existence (or not) of alternative modes, routes, destinations and timings.

Issues of time and capacity are further complicated by the fact that road and rail systems both have to cater for traffic that moves at different speeds. Fast and slow trains make use of the same network, but stopping trains reduce the paths available for high-speed services. Juggling timetables is thus a matter of juggling the value of different forms of service, and of handling the interests of competing

organisations and groups of customers. These situations are especially delicate in that they involve a mixture of network management and more specific forms of timing and provision.

From the consumers' point of view, railway systems are not 'always on'. Instead, train services are organised around well-defined operating schedules and fixed timetables. These have a double role, determining the timing of (met) demand, whilst also being developed in response to it. Since the scale and frequency of provision of trains, flights or bus services fluctuates, effective capacity varies during the day. These variations are not random but are instead related to, and part of more extensive societal rhythms, including office hours, school holidays and shop opening times.

In combination, these examples show that material arrangements, institutions and operating procedures are crucial for temporal fluctuations in supply and demand and for how these are accommodated and managed. Even more important, these systems of provision are integral to the timing of the practices they enable. This recursive aspect is routinely overlooked by policy makers or by utilities bent on building extra capacity or making better use of assets in order to meet need whenever and at whatever scale that might arise. In assuming that needs already exist, such approaches overlook questions about how demand is constituted, how expectations of service emerge and change, and how peaks and strains are interpreted and understood. This far, we have made much of the fact that material infrastructures and systems of provision are implicated in both the extent and the timing of demand. This is a necessary first step, but we need to go further if we are to explain daily and seasonal fluctuations in demand, and if we are to understand how peaks and troughs are formed.

Peaks and troughs in daily life and in demand

To address these questions we need to say more about why specific practices happen when they do, and about the forms of sequencing and synchronisation involved. Let us start with electricity. Peaks and troughs in electricity demand are a product of the aggregated switching on and off of the sum total of powered devices within the network or system in question. The rhythms of industrial and commercial electricity demand are not the same as those of the domestic sector, but it is the combination of these profiles that determines both the extent and duration of daily and seasonal peaks and troughs. In what follows, we focus on the ups and downs of domestic consumption on the grounds that this is crucial for the formation of

the 'highest' peaks in electricity demand which occur in the evening when people return home from work and prepare an evening meal.

In simple terms, peaks in domestic electricity consumption have to do with the number of appliances in use at any one moment and the total amount of power that such devices draw. Appliances and devices link to practices in ways that structure the timing of consumption. This works in different ways. Appliances like fridge freezers and wireless routers are on all the time; televisions, computers and printers are often on standby (Meier and Siderius, 2017); meanwhile heating and hot water systems separate the timing of demand from the timing of service delivery (Morley, 2017). In other situations, the timing of gas, water or electricity demand is directly tied to the timing of specific practices. This is the case when using a sewing machine or a gas hob, when taking an electric shower, or when working on a production line.

These different configurations of materiality, practice and timing combine to constitute the pressure on the network as a whole. However, this is not a one-way effect. The details of appliance design also have a bearing on the details of related practices, and when they happen. For example, the introduction of automatic washing machines transformed both the meaning of doing the laundry, the effort and time involved, and how this activity fitted into the daily or weekly routine (Spurling, 2018). Washing by hand or with a 'twin tub' used to be a task that required fairly constant attention in its own right: moving washing from one tub to another, draining, rinsing, loading and unloading etc. Automatic machines changed all this, turning washing into a task that could be woven into household schedules alongside other practices. In the past there have been deliberate attempts to generate load by electrifying practices that were likely to happen at specific times of day (Forty, 1986). However, domestic schedules are not simply defined by the technologies and material infrastructures on which they depend. They are also formed by patterns of sequencing and societal synchronisation. These features are simultaneously evident, but also disguised in electricity load profiles like the one illustrated in Figure 4.1.

It is fairly obvious that the dip in electricity demand between about midnight and 6.30am has to do with the fact that many people sleep at night. But what about the rest of the figure? Exactly what combinations of social practices lie behind the daily load profile; how do these come together to form periods of peak and off-peak demand and how do these differ across the year?

Matching data on time use, mobility and domestic energy use provides some clues as to what people are actually doing at different

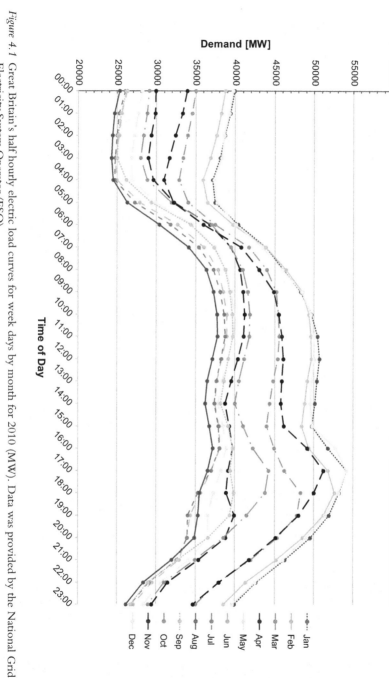

Figure 4.1 Great Britain's half hourly electric load curves for week days by month for 2010 (MW). Data was provided by the National Grid Electricity System Operator (ESO).

times of day and where they are. Amongst other things, disaggregating such data shows that the practices that lie behind evening peaks in electricity demand vary through the week. In the UK, the mixture of activities that characterise Monday evenings is not exactly the same as that recorded on Fridays, when (on average) people return home from work earlier and go to bed later than on other days of the week (Torriti, 2017). The weekends are different again, with distinctive time and demand profiles on Saturdays, when many shops are open, as compared with Sundays when some are closed. Analyses like these show that whilst some peaks in electricity or travel demand arise when many people do the same resource demanding activity at roughly the same time (e.g. meal times or commuting), others are the result of many people doing different but energy intensive practices at the same time. Travel peaks on Saturday mornings are of this more varied kind.

As well as revealing daily, weekly and also seasonal variations, data on time use and energy consumption show that in the UK, many people are regularly engaged in preparing and cooking an evening meal at some point between 5.00 and 7.00pm each day. Since activities relating to the evening meal account for at least some of the evening peak in electricity demand, it is useful to reflect on the origin of this arrangement and how it is sustained. Although people need to eat, social and cultural histories show that the number and the timing of meals is not a matter of human biology, or of personal preference alone (Yates and Warde, 2015). Not so long ago, in the UK, the main meal of the day was typically taken at midday rather than in the evening. In explaining why this is no longer the case, Southerton refers to the value placed on eating together as a family, the challenges of coordinating the schedules of different household members, and the significance of rising levels of women's employment (Southerton, 2009). These and other collective trends, including extended commuting times, help explain the shift from having the main meal in the middle of the day to having it at night.

But what about the detail: why eat at 5.30, or 6pm, or later? To some extent, the timing of dinner is a consequence of how this practice is positioned within a sequence of necessarily related events. These are likely to include travelling back from work and having time to prepare and cook a meal before eating it. Yet this is not a sequence that is fixed or set in stone. For example, in France, people also have their main meal in the evening, but tend to eat it rather later – a feature that is of some significance for when the peak in electricity demand occurs, and for what other activities happen between

returning home and going to bed. Time use data shows that using the internet, or watching TV are differently distributed across early and later evening periods, and that patterns in France and the UK are not the same, partly because of when the evening meal takes place, and how other activities are organised around it.

More abstractly, and thinking of the day, week, or year as a whole, peaks and troughs are not simply defined by the enactment of one practice or another, but by how activities are sequenced and scheduled in relation to each other. This is obvious if we think about the time and energy related consequences of the arrival of a new practice. Before television ownership became widespread, people in the UK did other things with the hours now devoted to watching TV: on average 3.2 hours a day (Ofcom, 2018). With the diffusion of televisions, and with the deliberate programming of prime time shows, different ways of spending time in the evening took hold. Watching television remains one of the most common evening pursuits, but there is some evidence to show that the potential to watch films and TV on the internet (including YouTube and Netflix), and to do so 'on demand' is eroding previously fixed schedules. Even so, patterns of viewing are not random or widely distributed: they are still fitted in alongside or as part of other practices, and still defined by collective institutional timings such as school hours or the working day.

In the UK, the detail of what people do across the day, and between week days and weekends also reflects seasonal variation in the weather and in hours of daylight and darkness. This is evident in annual load profiles of gas and electricity, as shown in Figure 4.2.

At this level of aggregation, seasonal variations in electricity demand are significant at around 30–40%, but small when compared with the annual swing in gas consumption. This does not simply reveal the seasonal need for heating: it is also an outcome of widespread reliance on gas rather than electricity or coal, linked to the diffusion of full central heating (Trentmann and Carlsson-Hyslop, 2018). In hot countries, and where air-conditioning has become the norm, electricity demand tends to peak in the summer. Whenever they occur, and whatever the fuel involved, seasonal fluctuations in demand do not simply depend on the outdoor climate alone. They are also consequences of changing ideas about what counts as a normal and a comfortable environment indoors. In other words it is the project of maintaining a steady 21 or 22°C indoors all year round that underpins some of these seasonal swings, not the weather as such (Shove et al., 2009a).

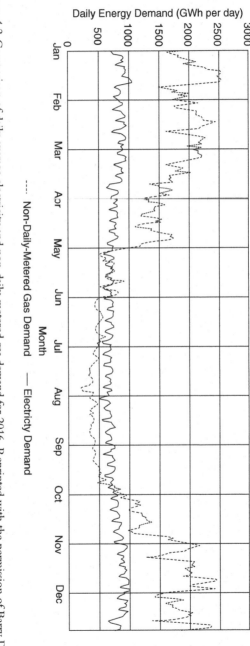

Figure 4.2 Comparison of daily average electricity and non-daily-metered gas demand for 2016. Reprinted with the permission of Barry D. Smith, EPSRC Centre for Doctoral Training in Energy Storage and its Applications.

This is an important point, but one that is systematically ironed out of most energy research and policy. Rather than tracking the ways in which interpretations of comfort have changed over the last century or so, energy analysts go to considerable lengths to exclude these trends. This works in various ways, some of which are buried in methods of representing demand. Since the amount of energy used for heating or cooling rises or falls depending on the temperature outside, identifying underlying trends depends on taking this factor into account, and on then taking it out of the equation. Methods of weather correction and seasonal adjustment are designed to do just that. There are different approaches, but one is to correct actual data on the energy consumed on any one day, and to adjust figures up or down to take account of how the weather on that particular day compares to a long-term average for the time of year. This technique supposes that indoor conditions remain the same from one year to the next. An alternative is to use what are known as 'degree days'. Heating or cooling degree days represent the number of days on which outdoor temperatures are higher or lower than a baseline temperature, and the extent of the difference. Designers and engineers refer to degree days when figuring out how much heating or cooling to install and how powerful this should be. Degree days are also used in establishing correction factors for energy data. In both roles, they have the important but also unacknowledged effect of reproducing and sometimes escalating interpretations of demand. As Rahman explains:

> In building energy management, the baseline temperature is the temperature that it needs to be outdoors so that no energy is required for heating the building ... Historically in the UK degree days have been calculated for a baseline temperature of 15.5° (e.g. by the Met Office). But this baseline is based on American dwellings data from the 1920s and seems rather implausible as a target temperature now.
>
> (Rahman, 2011: 24)

Resetting the baseline to 18 rather than 15.5°C changes estimates of how much energy is needed and of when this demand is expected to occur. If designers assume a baseline of 18°C that is what home infrastructures, and more extensive networks of supply will be designed to enable. When this convention becomes inscribed in daily life the heating season lengthens and the timing and extent of seasonal demand changes. This is an admittedly complicated example, but it shows how social conventions become embedded in energy research and analysis, in building design and ultimately in the seasonality of energy consumption.

There are other more obvious connections between seasonal fluctuations in demand and social practice. Holiday periods are a good example. Since total electricity demand is generally less when business and industry are closed and when people are at home, school vacations contribute to periodic dips in demand. According to Saddler, the impact of special occasions, like Christmas is more complicated, depending on a combination of factors including the day of the week on which 25 December falls, and the weather conditions on that day (Saddler, 2013). Other changes in demand simply reflect the fact that some people do different things, or do the same things differently, at different times of year. For example, clothes are more likely to be dried in a tumble dryer during the winter, and out on the clothesline when the weather is fine (Drysdale et al., 2015). Although there are seasonal and weather-related swings in practice, various commentators claim that people are becoming increasingly insulated from natural rhythms (Hitchings, 2010). This is important in that there are energy costs involved in maintaining standardised conditions and practices through the year.

As all these examples suggest, contemporary peaks and troughs are neither inevitable nor fixed. They are in various ways, consequences of how networks and systems of provision are designed, sized and operated, and of societal rhythms constituted by the sequencing and synchronisation of a plethora of social practices. Daily, weekly and seasonal patterns coexist, and also shape each other. These interdependent processes hang together, resulting in relatively regular, relatively predictable load profiles across many sectors.

In the last part of this chapter we discuss the significance of this practice-based account for current debates about flexibility, and for deliberate interventions to manage the timing of demand.

Flexibility in supply, in demand and in daily life

The traditional strategy of predicting future needs for energy and transport, and providing infrastructures and resources capable of meeting demand at any time and at almost any cost is incompatible with the goal of significant decarbonisation. One response is to look for ways of making better use of existing infrastructures and of doing so through the more efficient management of the relation between demand and supply. This is one reason why there is increasing interest in the concept of flexibility, especially in the electricity system (Carbon Trust, 2016). Another

is that modifying the supply mix to include more sources of renewable energy whilst also promoting the use of decarbonised electricity in place of fossil fuel threatens to put new strains on the system as a whole. This leads Ofgem (the regulator) to conclude that:

> Flexibility is increasingly central to this transforming system. Technologies and applications such as storage and demand response can help balance generation with demand and provide essential services to the grid. This can facilitate the deployment of weather-dependent renewables such as solar and wind, whilst enabling greater uptake of new types of demand such as electric transport.
>
> (HM Government, 2018: 3)

According to Ofgem there are different reasons why the supply–demand relation might need to 'flex' in the future. These include weather-related variations in power generation along with new profiles of demand linked to the emergence of practices like charging electric vehicles. However, the challenges of balancing the timing of supply and demand are not, of themselves, new. Histories of infrastructural development and use are often marked by short-term disruptions in supply such as power cuts, or by political and economic events that restrict the flow of oil or gas, and that call for adjustments in the extent and timing of demand (Chappells and Trentmann, 2018). Chappells and Trentmann's analysis of different forms of disruption suggests that far from being exceptional, balancing supply and demand, and doing so over the timescales of minutes, seconds, days or months is an ongoing process. Precisely how this is organised and exactly who and what is involved partly depends on the institutional arrangements and politics of the day.

On the supply side, methods of coping with fluctuations in demand include modulating generation, storage or substitution. For example, where resources (logs, oil, water) can be stored, temporary imbalances in the supply–demand relation can be handled by dipping into the reserve, or adding to it. However, not all resources or forms of service provision can be buffered in quite this way. The transport sector is intriguing in this respect. The capacity to access mobility, and to do so at a specific moment, depends on the availability of infrastructures and of relevant means of travel: bus, car, bicycle etc. Having a bicycle or a car ready and waiting on the forecourt enables people to move at a moment's notice. This is not so for those who have to plan their journeys to catch a train or a bus. Here, the means of access (the bus,

the train) is only briefly available at any one location: in this it is like a highly perishable commodity. On the other hand, if there is a really good service, missing a specific bus or train does not compromise the ability to complete the journey or to do so more or less on time. In this example, there are different ways of conceptualising provision, meaning that there are also multiple interpretations of what it means to ramp supply up or down, and how and by whom this might be done.

Classic forms of supply side flexibility involve modulating the supply of resources or services and/or making use of different forms of storage. Delivering the same services but by other means is another option. In the energy sector, fuel switching is one such example. Offering different modes of transport might be another. In both cases further questions arise about the meaning of equivalent service. It is true that central heating systems can run on gas or oil and that both fuels deliver the same kind of heat. But what if the switch is between using a bus or taking the tube? In some respects the service is the same – the destination is reached, but not at the same speed, not in the same way and not with the same level of reliability, however that might be interpreted. As these examples indicate, what look like supply side responses often have recursive implications for demand.

Further complexities arise when considering what are known as methods of 'demand side management' or 'demand side response'. Despite the name, demand side responses (DSR) do not always involve or require changes in the detail or the timing of what people do. In fact many forms of DSR are explicitly designed to modify load profiles without compromising 'consumer utility' (Drysdale et al., 2015: 282). In the electricity sector, what are known as flexibility providers (sometimes aggregators, sometimes large organisations) organise, deliver and sell decreases or increases in demand to optimise the efficient functioning of the network as a whole. Often this is a matter of temporarily switching off loads, for instance for industrial processes, or for cooling and heating, or deliberately turning up demand to take advantage of a glut of renewable energy, especially in the summer (National Grid, 2017).

Whilst these strategies are generally expected to operate behind the scenes, and to have no noticeable effect on the rhythms of daily life, other forms of demand side management imply more or less radical changes in consumer behaviour (Powells et al., 2014). In this context, there are ongoing debates about whether behaviour change, usually in the timing of activity, is best achieved 'via technological means or via economic incentives' (Smale et al., 2017: 134) such as peak time

pricing. There are also relevant discussions about whether certain groups of people have more flexibility to offer than others. For example, various authors conclude that those who are retired or without children have more opportunities to shift energy demanding activities to different times of day than those who work full time and/or have a young family (Torriti, 2015). There is also interest in establishing whether some practices are more flexible and thus more amenable to intervention than others. It is for this reason that Powells et al. report on the contrasting flexibilities associated with cooking, dining, laundry and washing dishes (Powells et al., 2014), as indicated by consumers' responses to a time sensitive tariff. Their results echo the conclusion that energy demand relating to the laundry is more flexible than that associated with cooking dinner (Anderson, 2016).

Whether they focus on the demand or the supply side, the responses outlined above are at odds with our practice-based account of resource consumption. One difference is in how the timing and scheduling of energy-demanding practices is conceptualised. For example, 'demand side' efforts to single out more and less flexible practices lose sight of the totality of what people do during the course of a day or a week, and of the overlapping forms of sequencing and synchronisation involved. Similarly whilst some individuals might seem to be more flexible than others, this is a consequence of socio-temporal arrangements that exist beyond the diary or schedule of any one person. Rather than conceptualising flexibility as a property of a specific practice or a person, we take it to be an outcome of the range of social practices enacted in society, and of the societal rhythms they engender.

By implication, more flexible societies would be those with fewer instances of synchronisation, and fewer collective patterns (e.g. fixed opening or working hours), along with more diverse practices and interpretations of service provision. Some argue that this implies greater individualisation and a fracturing of social ties that underpin collective as well as personal well-being (Putnam, 2001; Sennett, 2011; Durkheim and Lukes, 2013). Whether this is a good thing or not, more flexible societies are also likely to have more variable and perhaps more precisely controlled systems of provision as compared with infrastructures that build and also take advantage of synchronised and standardised economies of scale. Although they do not help explain how any of these features emerge or change, contemporary time use studies provide some empirical evidence of variation in the extent of societal synchronisation and, by implication, the extent to which socio-temporal patterns are shared. There are, for instance,

striking differences in how meal times are arranged in France and in Finland (Shove, 2009). In France, many people sit down to lunch at around midday but in Finland, time spent eating is more evenly distributed. This pattern has consequences for how other practices are coordinated and timed, meaning that there is, at least in theory, more scope for shifting specific activities to off-peak times in Finland than in France. If greater flexibility in the societal coordination of social practice underpins the scope for greater flexibility in balancing supply and demand, the next question is whether societal rhythms can be deliberately reconfigured with this ambition in mind.

From what has already been said, meal times, working hours and other institutionally timed events have an orchestrating effect on the sequencing and scheduling of related activities. For authors like Pred and Hägerstrand 'dominant projects' linked to work and family life generate coupling constraints that have effect on, and are also constituted by, the time-geographical paths that people take through the day (Pred, 1981; Ellegård and Svedin, 2012). In the present context, the key point is that these patterns are influenced by many forms of policy, for instance in areas such as health, employment or education. Such policies are rarely designed to have an impact on the timing of demand, but some are. In the Netherlands, school summer holidays start on different days in different regions of the country. This is the result of a policy that is quite explicitly intended to limit otherwise problematic forms of synchronisation: spreading the holiday season to alleviate peak pricing and periods of intense congestion.

Other policies including those that support flexible working hours, or working from home also enable a greater variety of social-temporal organisation. However, regulations and interventions of this kind do not necessarily result in the collective reconfiguring of energy-demanding practices, not least because so much more is involved. Although many organisations offer forms of flexible working, including opportunities for working from home, and although there is some evidence that different patterns are emerging at the edges of the working day, there is as yet no dramatic swing away from the 9.00am–5.00pm regime. Changes that have occurred over the last 40 years, including a slight lengthening of the working day, and the consolidation of commuting 'into a smaller number of journeys' (Le Vine et al., 2017), appear to have little or nothing to do with the ambition of reducing congestion (Figure 4.3).

The slow and patchy pace of change is not surprising when we remember that working practices are held in place by so many other routines and conventions (meal times, school times) and by established

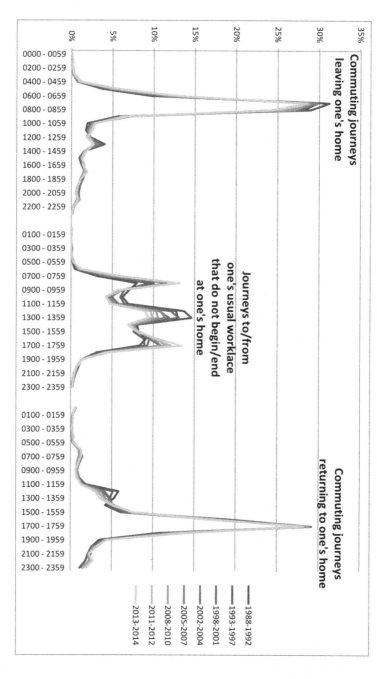

Figure 4.3 Journeys to/from one's usual workplace by hour-of-day (AM peak period). Panels show journeys from one's home (L), to one's home (R), and neither beginning nor ending at one's home (centre). Source: Le Vine et al., 2017. Licensed under crown copyright for re-use with attribution, Open Government Licence v3.0.

infrastructures, including house design, working space etc. This does not diminish the point that past and present policy makers influence the social-temporal ordering of society: they clearly do whether through investing in new networks and systems of provision, or by modifying institutional and therefore also domestic routines. Rather, the point is that the effect of these interventions is mediated by existing arrangements and by other apparently unrelated, but also critical 'projects' and 'coupling constraints' (Pred, 1981).

On the face of it, these observations lead us to conclude that deliberately modifying the timing of energy demand depends on modifying the sequencing and scheduling of social practices. This is an important contribution, but our analysis of flexibility, and of how it is defined and understood points to another even more critical insight. As we explain below, contemporary methods of calculating and representing flexibility rest on a bedrock of assumptions about the fixity of demand.

In the electricity sector, quantitative estimates of flexibility represent the scale of the potential to turn supply or demand up or down to 'help balance the system'. The Association for Decentralised Energy, which takes this approach, claims that there are 'many hundreds of megawatts of business flexibility' that could be exploited, and that the UK could deliver 9.8 gigawatts of flexibility by 2020 (Association for Decentralised Energy, 2016). Methods of arriving at figures like these take account of factors such as 'load interruptibility, time of use flexibility and the willingness of consumers to cede control over their appliances' (Drysdale et al., 2015: 287). As this list confirms, these estimates focus on variously flexible means of delivering fixed levels of service, and not on the potential for flexing the very meaning and purpose of demand.

To some extent this makes sense. After all, it would be exceptionally challenging to think about how daily and seasonal rhythms could change, or to imagine a future world in which domestic and industrial practices were significantly different from those with which we are familiar today. But in taking present levels of service for granted methods of estimating, valuing and conceptualising flexibility overlook the fact that demand, in the more fundamental sense of what energy is used for in society, is itself dynamic. More than that, these working assumptions and understandings have the further, also paradoxical consequence of perpetuating contemporary conventions and societal rhythms along with the inflexibilities these contain. It is easy to understand why questions about what energy is for are so regularly overlooked, but failure to engage with this topic represents a major

blind spot that severely restricts discussion of routes towards decarbonisation that involve somewhat, or perhaps significantly different societal rhythms. In the next chapter we take stock of the practical and political implications of this paradox, and of opportunities for intervening to steer both the extent and also the timing of demand.

Before moving on to a discussion of intervention it is useful to take stock of the ways in which demand is temporally embedded. In this chapter we have argued that peaks and troughs in demand are outcomes of the totality of social practices enacted in society: they are not expressions of individual choice, nor are they products of material culture, institutional design or engineering alone. In detail, peaks are defined by troughs just as much as troughs are defined by peaks. Put differently, there is a relation between how activities are bunched and spread, and something of a zero sum aspect to consider as well. Doing certain practices prevents others happening at the same time. In addition, and as we have repeatedly observed, forms of supply and demand constitute each other. This has consequences for the details of timing and synchronisation. For example, infrastructures assume certain combinations of practices and in enabling them to take place when they do, they reproduce the sequencing and scheduling, as well as the overall scale of demand.

The ways in which this works are neither simple or direct. Aspects of timing, duration and frequency are to a degree shaped by material arrangements, encompassing but extending beyond the design of specific appliances and including more extensive networks and systems of provision. How these material arrangements work out in practice depends on how they are organised and operated and on the institutions and social conventions involved. Together these arrangements define not only the positioning and extent of peaks and troughs but also the potential to cope with fluctuations in demand or in supply, for instance through forms of storage and substitution or by rescheduling energy demanding practices.

Contemporary policies and forms of intervention that focus on the timing of consumption tend to conceptualise supply and demand as separate spheres. They consequently overlook the extent to which infrastructures and systems of provision are implicated in what people do, the practices they enact and when these occur. In exploring these themes, and in discussing flexibility as societal issue bound up with collective patterns of timing and synchronisation, we have opened the way for more adventurous approaches, including those that entail changes in everyday practice, and thus in the extent of demand and in when it occurs.

5   Shaping and steering demand

So far we have discussed a series of linked propositions, using the first two chapters to suggest that demand is derived from social practices, that it is made, that it changes and that it does not simply exist as something that needs to be met by market actors or by policy makers. Chapters 3 and 4, on matters and temporalities of demand, went into further detail showing that demands for energy and for mobility are both materially embedded and temporally unfolding. In this chapter we recognise that the extent and the timing of demand is influenced by a wide range of government and organisational policy making, whether that is the intention or not. More than that, we argue that rather than taking demand for granted, or conceptualising it as a non-negotiable 'need', policy makers charged with carbon reduction could and should recognise their part in constituting demand and treat demand reduction as a legitimate arena for intervention.

In practice, and as we are well aware, the prospect of establishing an effective demand reduction policy is challenging on a number of fronts. For a start, there are limits to what can be achieved through policy intervention alone: simple cause and effect relationships rarely exist and the connections between so-called policy levers and outcomes are never straightforward. This is partly because the social practices on which demand depends are not simply the result of exogenous policy prescriptions or of political and economic systems, nor are they expressions of individual agency and choice. Instead, relevant practices evolve endogenously; they are multiply interlinked, and they are shaped and defined by their various and connected histories. However, this does not mean that policy is unimportant.

The claim that all sorts of policies influence demand, and that they do so in a range of different ways might lead some to conclude that since everything could be construed as a form of 'demand policy', the concept is too nebulous to deal with. We argue otherwise. After all,

many factors influence equally complex problems such as the extent of poverty or national levels of educational attainment but this does not prevent policy makers from seeking to address these issues. Usually, the method is to identify the most important aspects which policy *can* affect and target action there. In the case of carbon reduction, this begs the question: which policies are most important in shaping future demand?

From what we have said already, it seems clear that policies that are explicitly to do with energy (often focusing on efficiency) are unlikely to be those that have greatest impact on energy demanding practices. On the other hand, the fact that many policies are of consequence for energy and mobility demand does not mean that all are equally important or equally amenable to action. To illustrate both the possibilities and the challenges of developing policies to reduce demand we focus on an example from the transport sector.

Transport is a particularly relevant case in that the movement of people and goods accounts for 25% of global energy consumption (EIA, 2016). It is a sector in which the outcomes of highly synchronised demand are typically more visible (congestion, overcrowding, pollution) than they are for something like electricity. It is also a sector that recognises, at least in part, that the demand for mobility is *derived* from the need to participate in activities as part of everyday life. As such, one might expect that discussions of how demand is made, and of how the timing of mobility is embedded in the rhythm of society would be central. However, as the case study demonstrates, the dominant framing of travel behaviour and the associated politics of choice are such that deliberate strategies of demand reduction are few and far between. This then is a setting in which there is scope for reframing and rethinking what demand reduction might involve. In the following sections we set out some examples of how this could be done before broadening back out to the implications of our analysis for other policy areas.

The role of policy in steering and shaping demand

Policy, Birkland (2016: 8) argues is not straightforward to define. He suggests that some key elements of public policy are that 'it is made in response to some sort of problem' and that it is 'made by governments' but 'interpreted and implemented by public and private actors'. As Dye (2012) argues, policy is also what governments choose not to do as well as the actions they take. If policy is treated as being made in response to some kind of problem then, by definition, it is seeking

to resolve or diminish the problem, whether or not it is successful in so doing. For example, poverty could be treated by expanding welfare payments or through new housing policies. Noise pollution could be tackled by regulating technology and/or limiting the times of day at which certain activities can happen. Water shortages can be dealt with through expanding supply, changing prices, fixing leakages or rationing (e.g. hosepipe bans). Each of these are policy actions that could be deployed to change demand in some way or another. Taking the water shortage example:

- Expanding supply normalises additional levels of water demand which may contribute to future shortages.
- Changing prices takes a position that there is more demand than is deemed desirable from an economic perspective (see Chapter 2) and that some 'low value' usage can be discouraged if consumers are faced with a better cost structure.
- Fixing leakages is a different route to increasing supply but does not involve expanding the existing supply network. In reality such a solution will likely exist alongside either or both of the above.
- Rationing supply is often seen as an emergency response only, although there have been droughts affecting water supply in parts of the UK for 13 of the past 40 years (EA, 2017). This response reflects the view that some aspects of water demand are negotiable and that the costs of maintaining 'normal' levels of supply throughout all rainfall scenarios are prohibitive.

Water supply is in many ways an obvious example in which demand is visible, especially when supplies are low, and which matters in public policy terms, at least some of the time. In almost all the possible responses outlined above, existing water consumption is a given and any underlying causes of increasing demand, such as increases in the use of appliances like dishwashers and washing machines or more frequent showering, are not addressed. In this case, as in many others, policy makers can be simultaneously focused on questions relating to demand, especially on how to deliver sufficient supply, whilst being blind to questions about how that demand came to be in the first instance.

If we focus on this latter question, it is plain that demands for energy and for mobility are often strongly influenced by policies that are not explicitly to do with either of these topics. Royston et al. (2018) expand on this notion, listing a whole series of policies from the provision of free schools, through the indexing of pensions to

inflation to the centralisation of specialist medical services in some hospitals, claiming that each has consequences for energy demand. As cited by Royston et al. (2018), the provision of free schools is about *how* capacity for new school places is managed. Debate about specialist services is about *where* care should be provided to maximise health outcomes. Linking state pensions to inflation relates to a balance of affordability and some notion of what a reasonable level of income might be for a retiree. These are all reasonable goals, but the fact that these policies have a recursive impact on the demand for energy or mobility is not, in itself, exploited as a means of demand management.

There is undoubtedly scope for actively using non-energy policies to reduce demand for energy and mobility, and to do so at source. Equally, and as we have already said, demand is not defined by policy alone. For example, in the area of health care, new technologies can generate demand for new forms of treatment, resulting in energy and transport demands that did not exist before. Going one step back, the prevalence of certain diseases and thus the need for treatment might also increase due to changing patterns of activity and diet. The point is that although policies of many kinds can and do have consequences for energy demand, this is mediated by existing complexes of social practices, infrastructures and technologies. It is for this reason that we talk of 'steering' and 'shaping' demand, recognising that policy is but one influence on what are almost always complex, interdependent and historically defined trajectories.

A further observation is that different forms and types of policy combine and interact, with results that are not always easy to anticipate or predict. For example, some policy interventions (e.g. introducing student fees in Higher Education) are not expected to increase energy demand, but they have had this effect. Policy makers are not blind to these kinds of interaction, and some are quite clearly in view. In health care there is, for instance, explicit interest in how telemedicine or the provision of local services via pharmacies might reduce travel demand. However, it is important to remember that policies are typically organised around specific problems, such as optimising health care. If this is the *primary* ambition, the effects on transport or energy demand figure as *secondary* considerations, and are treated as such.

This helps explain why energy demand reduction is such a challenging topic. Since demand is 'an outcome of what people and their machines do in their homes, at work, in leisure time, and in moving around' it is 'powerfully shaped by, among other things, a wide range of policy priorities and processes' (Royston et al., 2018: 128). Whilst some of these policies are directly to do with energy and

its consumption, most are not. Royston et al. define these as 'non-energy policies' – meaning that they are policies that shape demand, but that do not explicitly take this into account. A review of 576 academic and grey literature studies of the impacts of non-energy policy (e.g. in agriculture, communications, culture) on energy systems found that although there was some interest in mattes of energy supply (e.g. links between land use and biofuels), changes in demand and in what energy is used for in society were almost entirely invisible (Cox et al., 2016).

As we have already pointed out, not all non-energy policies are relevant for energy demand, but some can be exceptionally important. Recent research has shown how treatment targets and discharge routines influence energy use in the health service (Blue, 2018); how welfare reform and housing provision for vulnerable families defines the energy they use (Butler et al., 2018) and how the introduction of undergraduate fees has changed building programmes and operational practices in UK higher education in ways that increase energy consumption across the sector as a whole (Royston et al., 2018). These studies suggest that the tendency to overlook the extent to which non-energy policies shape energy demand is not systematic or evidence-based. Often it is simply the result of positioning problems of energy demand as problems that are central for energy policy, but not for any other field.

Royston et al. (2018) go further, arguing that if this approach were to be reversed it might be possible to steer energy and mobility demand by identifying and making use of many different points of policy intervention. As they explain, legislation, budget allocation and strategic direction, for example through target setting, constitute diverse but relevant means of shaping policy 'on the ground'. Similarly, technical standard setting, which can occur within or outside the legislative process, along with guidelines and conventions of good practice help establish and normalise particular ways of doing things (see Faulconbridge et al., 2017 on the relationship between the ratcheting up of energy demand in the commercial office market and quality standards). These 'harder' and 'softer' forms of policy combine in the specification and also the enforcement or non-enforcement of rules (Lipsky, 2010), and in the actions of those responsible for the day-to-day provision of all manner of different services.

There are different interpretations of what this means for those seeking to use non-energy policies to reduce demand. The fact that policies that matter for energy demand are not made in isolation, or in any one specific forum, or by any one set of stakeholders (Rhodes, 2007; Bache

and Flinders, 2004), might be seen either as a confounding factor, or as an indication of potential and opportunity. In the next section we say more about what it might take to capitalise on the potential to reconfigure demand and to do so across a range of policy areas.

Reconfiguring demand for energy and mobility

The previous section outlined the importance and pervasiveness of non-energy policy in shaping demand for energy. Having established that there is a wide-ranging potential for policy making and processes of policy enactment to reduce energy demand we now focus on how this might be realised in practice.

One obvious implication is that there would need to be closer interaction between energy and non-energy policy makers of all kinds. This kind of 'joining up' is notoriously difficult, even in relation to apparently simple challenges like those of integrating policies on transport and land-use planning or between public health and hospital care (Stead, 2008).

Second, and building on the observation that policy making does not happen in the abstract, it would be important to consider the long-term as well as the short-term consequences of a range of non-energy policies. For example, ideas about what count as 'normal' comfortable conditions indoors have become established and embedded over a period of decades, and are now held in place not by one but by a range of policies along with informal standards and conventions. In combination, these justify patterns of heating and cooling that are set to dominate aspects of energy demand for years to come. In other words, and as Pollitt (2008) suggests, it is a mistake to focus on short-term policy outcomes and processes and to overlook the ways in which the effects of legislation and policy making unfold over time. In practice, this argues for at least paying attention to the ways in which contemporary policies amplify or depart from previous traditions, and how they shape future pathways and possibilities.

In making the case for longer term and more wide ranging forms of demand-reduction policy, we might be accused of failing to appreciate the challenges of resolving 'wicked problems' (Rittel and Webber, 1973) and of engendering horizontal co-ordination across policy areas (Christensen and Lægreid, 2007). One response is to explain that whether they recognise it or not, many different policy makers *already* have an impact on both the extent and the timing of energy and mobility demand. In our view, this does not necessarily result in a form of paralysis (Aucoin, 1990). Instead, and with careful

scrutiny and analysis, it should be possible to distinguish between more and less influential forms of non-energy policy, and focus on those that have most effect, now and in the future. As we have already noticed, the problem of systematic energy demand reduction is not a problem that is unique in its complexity or in its cross-cutting nature. The relationship between diverse policy actions and job creation or economic growth is at least as long running, complex, multi-headed and uncertain as is the construction and reduction of energy demand (Mullen and Marsden, 2015). In conclusion, public policy could have focused on non-energy policy as a means of demand reduction but, to date, it has not.

Changing this state of affairs depends on fundamentally changing the way in which the problem of demand reduction is defined and understood. Howlett et al. (2009: 92) argue; 'The manner and form in which problems are recognized, if they are recognized at all, are important determinants of whether, and how, they will ultimately be addressed by policy-makers.' In their seminal discussions of problem definition Rochefort and Cobb (1993: 15) argue that 'the function of problem definition is at once to explain, to describe, to recommend, and, above all, to persuade.'

In this context, the dominant conceptualisations of demand described in Chapter 2 matter significantly. As we explained, demand is usually thought to arise from consumers' desires, some of which are taken to be universal, non-negotiated matters of fact. In addition, and especially in economics, questions of demand are tied to questions of supply, and to a raft of related ideas about the part that pricing structures play in indicating and also managing tensions between the two. Given that demand is framed in these terms, questions about the changing characteristics of demand; its material and temporal qualities and its origins and histories are simply not on the agenda. And since these features are *not* seen to be part of the problem, paying attention to them is equally *not* part of the solution.

Instead, current approaches reproduce problem definitions and related solutions that are consistent with narratives in which the state provides or facilitates systems of provision and then encourages appropriate use by enabling consumers to make better choices for themselves (see Barr and Prillwitz, 2014 for a critique of 'smarter choices' in UK transport policy). This is important in that such narratives 'influence the shaping of laws, regulations, allocation decisions, institutional mechanisms, sanctions, incentives, procedures, and patterns of

behavior that determine what policies actually mean in action' (Schön and Rein, 1994: 32).

Challenging this paradigm depends on two linked moves: one is to acknowledge that the absolute volume of demand for energy (or whatever resource) is some way unattainable or undesirable. We see the next decade as one in which the realities of the need for rapid decarbonisation might take hold, alongside a realisation that existing individualised and efficiency-led demand policy are insufficient and potentially counterproductive (see CCC, 2018 for evidence of these tensions). Second, and even if the policy problem of demand reduction were to be repositioned in this way, there is still the question of how demand is itself conceptualised, and thus what possible responses and solutions might be in view.

In the literature on policy change, it is widely recognised that policy narratives are needed to legitimise and justify actions. Boswell et al. suggest that 'A narrative must meet certain basic conditions of consistency, coherence, and plausibility. Above all, it must "fit" with available facts about the case.' Narratives, they argue, also need 'to be understandable, compelling and sufficiently plausible for the actors in question' (Boswell et al., 2011: 6). On the face of it, generating a narrative capable of underpinning wide ranging, joined up interventions and of mobilising a range of non-energy policies to reduce demand might represent a step too far. On the other hand, and as we have said already, policy is about what governments choose to do, and what they do not do as well.

By failing to engage with non-energy policy and its impacts on demand policy makers have a position on this topic – just one that is chaotic and inconsistent and that has a net impact of reinforcing more energy demanding practices across society. Some aspects of a more coherent demand reduction policy (across non-energy areas) may indeed prove either too insignificant or too difficult to develop or enact in ways that are both worthwhile and effective. But this cannot be a universal truth, hence our argument is that there is a need to identify those places where action *could* be taken and to change the framing of the demand discussion so that these become legitimate and perhaps essential sites for intervention.

In the next section we explore this proposition further through an in-depth analysis of the treatment of travel demand in transport policy. Whilst each sector or issue will have a number of distinctive features, our goal is to use the transport case study to underline generic arguments that are relevant to other areas of policy and intervention.

Rethinking policy: a case study of travel demand

Demand has been a key feature of transport policy in the UK since the Second World War in particular as the growth of motorised traffic and decline in rail use simultaneously unfolded. As Marsden (2019) shows, actual and anticipated changes in demand have sometimes been mobilised as an argument for expanding investment programmes (e.g. the Roads for Prosperity programme of the 1990s) or for reducing them (e.g. Beeching's review of the rail network in the 1960s). Either way, observed or expected levels of demand are taken as a given, with attention then focusing on their positive and negative impacts (Reardon and Marsden, 2020).

One method of learning how travel demand is categorised and measured is to review the data and the methods that are used in making future projections. In what follows we examine the UK Department for Transport's road traffic estimates and forecasts to identify narratives and assumptions about current and future demand. Road transport in the UK currently comprises 24% of all national greenhouse gas emissions (CCC, 2018), and is almost entirely fossil fuel based although there is an anticipation that this could, or rather must, shift to electricity or clean hydrogen in the next couple of decades. In addition, transport is one of many sectors in which projections have been made for planning purposes over almost 50 years (this is also the case for electricity (BEIS, 2019), water (Wallingford, 2015) and health care (Appleby, 2013)).

The demand for travel can be measured in different ways, but one of the most common methods is to collect data on the number of trips made and the distance travelled for a range of different activities. In the UK, modelling future demand depends on estimating how such trips and distances might change by estimating changes in fuel prices, population, income and car ownership. As Figure 5.1 shows, the decades following the Second World War saw significant growth in car-based travel. The adoption of the car was one of the defining social changes of the latter part of the last century, so much so that the associated reordering of land-uses and social practices have been said to constitute a 'system of automobility' (Urry, 2004).

Research has shown that during this period the combination of data on fuel prices, GDP (as a proxy for income), changing car ownership and population was able to generate reasonably accurate estimates of growth in car miles driven (Bastian et al., 2016). Errors between forecast levels and actual outturns have been largely attributed to fluctuations in the economy or population levels which had not been

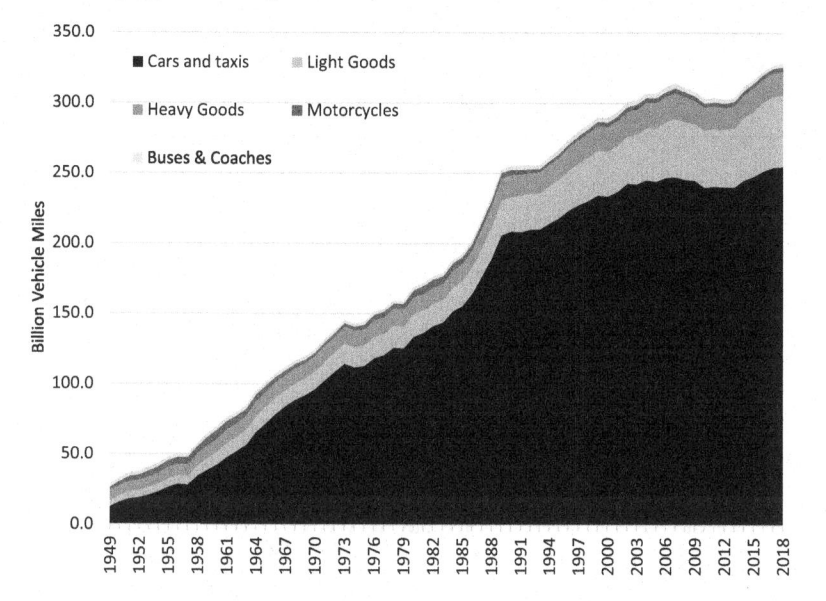

Figure 5.1 Growth in motorised traffic in Great Britain 1949 to 2016. Data source: Transport Statistics Great Britain.

foreseen. Over time, the strength of these relationships has changed and the Department for Transport (DfT) recalibrates its model every three or so years (DfT, 2018b). In September 2018 it released its latest forecasts which include a 'Reference Case' comprising a scenario of national mid-range population growth, oil price estimates and GDP forecasts, and variants on this. Figure 5.2 shows the growth implied by higher and lower estimates of GDP coupled with lower and higher fuel prices, these being the core parameters which have driven previous forecasts.

The Department for Transport's 2018 forecasts suggest that road traffic will grow by around as much in the coming 35 years as it did in the last 35 years. This is not surprising in that although relationships underlying the projections have been recalibrated, in all other respects the core parameters remain fundamentally unchanged. In addition, the Department for Transport only considers the consequences of policies and projects to which it is already committed, which means that the imagined social context for travel is largely fixed (2018b). Whilst these forecasts are only intended to inform future investment by the Department for Transport they are nonetheless hugely important in sustaining the narratives of growth around which the need for yet more infrastructure revolves. This picture of the future is, if anything, even

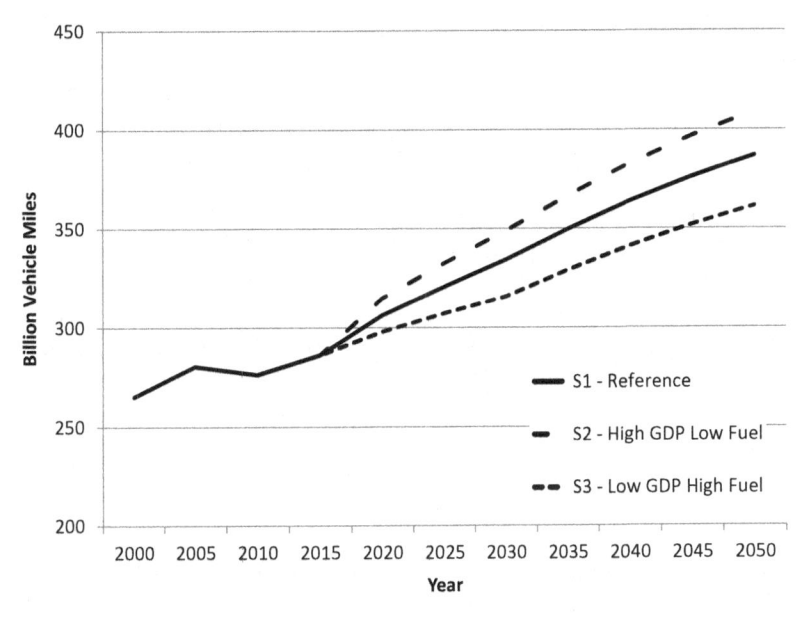

Figure 5.2 2018 Road traffic forecasts in England and Wales – Scenarios 1 to 3.
Data source: DfT, 2018b.

more clearly illustrated by the images of queues and slow moving traffic, which the forecast document also includes.

Current projections of *increasing* travel demand rest on a series of linked assumptions. One is that relationships between fuel price, income, car ownership and population are the critical determinants of travel and will remain so in the years ahead. This is not necessarily the case. The Commission on Travel Demand has explored recent changes in the underlying nature of travel demand rather than focusing on the headline miles driven (Marsden et al., 2018). From this work, it appears that per capita, people have been travelling fewer miles and making fewer trips for the past 25 years. More recently, people have also been travelling for fewer hours per year. This is true across almost all activity types (excluding education and escort education trips). It appears that younger people (defined here as under 30s) in particular are travelling less than previous younger generations whilst older generations have been driving more as a cohort that grew up with rising car ownership and use reaches retirement. It is only population growth, and a rise in light van traffic, which has produced traffic growth at an aggregate level. Since 1996 the population has increased by 13% and so has road traffic, although the one to

one mapping is coincidental. So, even if at an aggregate scale, the previous relationships have reproduced plausible estimates of traffic levels, the social practices underpinning this are significantly different from those that were assumed when 2020 was the forecast end date.

Second, if one were to take seriously the projections to 2050 it would be necessary to believe that the conditions that defined car growth in the latter half of the last century would still hold in the future. This seems to require several huge leaps of faith. For example, the suburban spread and the lengthening of the commute which accompanied the adoption of the car has already happened and cannot happen again. The expansion of the female labour market, although still unfolding, has already changed and the gap between males and females in the distances driven has closed: again these changes cannot happen twice.

Third, the projections do not allow for social changes which did not figure in the previous period because by definition, they are not included in current data and coefficients. For example, in the 1980's forecasts there was no anticipation of the advent of the internet. No allowance was made for home working or for the rise of internet shopping. These effects have still not finished playing out but as we have explained, the forecasting tool privileges historic economic drivers of change. In fact, it is only possible to represent new effects by modifying the coefficients from the past.

Even more important, the whole forecasting process does not allow for the imagination of futures in which demand is lower, or for the roles of policy in shaping these outcomes. This is because the Department for Transport insists that the forecasts are 'policy free' (DfT, 2018b). Such a position is intellectually moribund in that the model and the coefficients underpinning it plainly assume that the kind of policy environment that allowed for previous growth will continue in the years ahead.

There is some recognition now emerging that different demand futures could transpire. Both in 2015 and 2018 the Department for Transport has run a single scenario in which *the reduction* in trip rates observed over the past 20 years (16%) has been presumed to continue, rather than halting as in all other scenarios. This leads to a halving of estimated growth in traffic relative to the reference case by 2050, equivalent to almost one billion vehicle miles over the course of the forecast period. It is interesting to ask why presuming that trip rates will continue to decline is apparently less likely (expressed through its appearance in only one scenario) than presuming that this trend will be reversed in 2018 as is the case in all of the other scenarios. This,

we argue, is part of the performative politics of forecasting which we discuss further below. The lack of any account of the shifting practices to which mobility is linked allows changing trip patterns to be treated as an exogenous variable similar to fuel price, a *what if* rather than a *part of* ongoing processes of social change.

How does this matter for policy and for the future of demand? In the 1980s the forecasts were used to support a policy of major road expansion where the aim was to provide the capacity necessary to meet expected growth in demand, an approach known as 'predict and provide' (Owens, 1995). The roads programme of the late 1980s was abandoned due to protests over the environmental damage that resulted, the high costs involved and evidence that it would in any case fail to keep pace with demand (Goodwin et al., 1991). The period between the mid 1990s and the recession in 2007/08 was marked by a shift towards a strategy that involved deciding how much traffic growth to try to accommodate and how best to manage that growth (Goulden et al., 2014). Since 2010, there has been a shift back to investing in road (and rail) networks as a means of 'supporting a dynamic, modern economy' (HM Treasury, 2011: 5). Transport is the largest sector in the UK's planned infrastructure spending pipeline, with £54.9 billion committed to be spent by 2021 (IPA, 2018). Rather than being based on new or changing evidence these swings are instead indicative of shifts in political preference and discourse (Docherty et al., 2018).

Past and present forecasts feed into these narratives, and are real in their effects. As we know, expanding capacity on transport networks induces additional traffic (Hills, 1996). Putting additional infrastructure into our transport networks will open up new areas of land for housing and other kinds of development and if this land does not have accessible local services and good public transport links then it will contribute to rising traffic levels (Tennøy et al., 2017). Similarly, building new houses in inaccessible areas and designing them to accommodate increasing traffic (based on projections from the 1990s) is likely to facilitate, if not require growing demand for car travel (TfNH, 2018). If extra infrastructure is seen as critical to economic growth and if the provision of extra infrastructure supports the growth of traffic then the future is likely to be one in which people travel more.

This reasoning, the methods of forecasting and estimating associated with it and the forms of investment that follow do not create an environment that is conducive to thinking about trends and sets of policies that might actively reduce demand. However, the next three

examples show how changing technologies and social practices are impacting demand and they give a sense of what might be involved in fostering different kinds of futures, should policy be open to that.

- Employment trends: Despite there being evidence that there are more people in work than ever, fewer commuting trips are being made than in the 1990s (Le Vine et al., 2017). How will work evolve, what will the relationship be between transportation and work and which policies might influence this?
- Travel of younger people: Under 30s have been travelling less in the past two decades. Research has shown that this is largely a result of changes outside the transport sector such as delayed life course events like having children later, staying longer in education and the emergence of new patterns of housing and ways of living (Chatterjee et al., 2018). Might this continue and how much have wider government policies impacted on this (e.g. widening participation in higher education)?
- Retail trends: Internet shopping now makes up 17% of all retail sales by value and has been increasing fairly consistently. This trend is now impacting on the viability of traditional high street retailing in a significant way (Singleton et al., 2016). Is this inevitable? How might high streets be reshaped (e.g. through planning or regulation of delivery charges or consolidation of package consignments)? What could be done with the empty shops?

It is entirely plausible to design policies which support rising demand but there are also opportunities to generate conditions in which travel demand falls. To use the steering analogy from the chapter title, the tiller could be turned to encourage travel in one of two directions. What we can see from the transport case is that there is a highly institutionalised approach to thinking about demand. The tiller on road traffic has only ever been used to steer in one direction (growth), with judgements then being made about how far to turn it. Over time, the currents have shifted and become more favourable to a direction of declining travel, at least on a per capita basis, but because these underlying flows (which we could see as changes in social practices) are not part of contemporary policy analysis, they have little or no effect on decisions about infrastructure and investment.

To broaden the argument beyond transport, reliance on a strongly traditional set of tools for understanding demand, based on necessarily limited economic paradigms reflects the dominance of highly technical disciplines (Marsden and McDonald, 2019). In this context, more

demand is generally seen to be a good thing, with less demand being potentially problematic since it implies a loss in welfare compared to current conditions. The institutions, methods and discourses that have grown up around these interpretations have significant path dependencies and are difficult to penetrate with new evidence. However, even where evidence can be brought to bear there remains a separate, albeit connected, political narrative which associates increased demand with economic growth and that links measures which stimulate demand to pro-growth narratives. In such settings, policies and measures that imagine or work towards demand reduction struggle to compete.

Such a prognosis might seem bleak. An overarching policy mindset which is fully engaged with the ambition of substantial demand reduction would be favourable to a wide range of new interventions. However, since demands for energy and for travel are built up over time and through a combination of policy, infrastructure, technology and shifts in social practice, such interlocking arrangements can also be reconstructed without one overarching narrative. Indeed, to require a single narrative would be to fall foul of our own observations that policy is only part of the picture and there is no one controlling agent.

This chapter has explored the fifth of our propositions, that demand is moderated and modulated, deliberately or otherwise, via many forms of policy and governance. Along the way, we have demonstrated this to be the case across a range of policy fields and we have considered some of the challenges involved in bringing such a set of insights to bear on the existing policy system. Rather than concluding that this is all too difficult for policy makers we find that it is more a matter of which questions are in focus and which are not. We caution against dominant cause and effect models because they will not capture the complexities and interactions we establish in Chapters 2, 3 and 4. Policy intervention in demand is inevitably uncertain and risky, but it is also happening all the time. Within this field there is scope for a more systematic and targeted approach. It is on this basis that we suggest that where evidence points to the emergence of less demanding practices it is used to challenge the status quo in policy design. By these means, the demand implications of a full range of transport, energy and non-energy policies could be recognised and on occasion mobilised in pursuit of demand reduction. At a minimum, such approaches would allow policy makers to identify and actively promote conditions and strategies that support and accelerate the emergence of social arrangements and practices that serve to reduce, rather than increase demand.

6 Demand

A distinctive approach

Despite the bold claim of the title, *Conceptualising Demand: a distinctive approach to consumption and practice*, this book weaves between ideas and arguments that are not, of themselves, especially distinctive or new. In working through our core propositions – that demand is an outcome of social practice; that demand is made, not simply met; that demand is materially embedded and temporally unfolding and that demand can be steered via many forms of policy – we build on arguments that have been made before. Aspects of our core agenda also have a history. Some 20 years ago, Wilhite et. al claimed that 'the nature and causes of "energy demand" have been oversimplified, reduced or ignored in the community of energy research and policy' (Wilhite et al., 2000: 109). In that paper Wilhite et al. also made the case for 'recasting the demand for energy, and the things which use energy, as a social demand, dependent not just on prices and degree of consumer awareness, but also on social norms and a network of social institutions' (2000: 109). In the intervening years much has happened in social theory and in energy and transport studies yet there is still something of a split. The propositions around which this book is organised tap in to increasingly lively streams of research and enquiry, but the problems to which they relate and the paradigms we critique appear to be as stubbornly entrenched as ever. It is in relation to these still dominant positions that ours counts as a distinctive approach.

There will always be different ways of defining demand and our aim is not to provide a single, definitive, all-encompassing account. In writing this book, and in persuading people to read it, our more modest ambition is to encourage those who take demand for granted, who take it to be the logical the partner to supply, or an expression of consumers' needs and desires to pause and to think again about the assumptions they make.

Although we have discussed them separately, the propositions developed in the preceding chapters cannot be read in isolation. All relate to the project of conceptualising demand, whether for resources, goods or services, as an outcome of historical and situated connections between things (from road networks, and power grids to fridges and foods) and practices of consumption and provision. In working with these ideas, and in working them through with reference to examples from the energy and transport sectors, we have shown what such an approach involves.

The result is not a simple guide to action, but it is a kind of manifesto. In brief, we contend that questions about the constitution and the transformation of demand are crucial, not only for climate change policy but for other sectors too. More than that, we claim that such questions could and should be topics of concerted analysis and intervention. In contrast to those who treat supply and demand as abstract concepts that have a role within similarly abstract theories of market exchange, we focus on matters of substance and detail. In taking this position we ask what it is that is 'demanded' at different points in time, and how these demands are themselves embedded in the changing practices and arrangements of which societies are composed.

This leads us to suggest that systems and infrastructures of provision and supply are integral to the making and shaping of the many practices on which demand depends. We have, for example, shown that food provisioning and refrigeration are interwoven with the details of everyday diet. Similarly, we have argued that new forms of shopping are reproduced and enacted in the practices of providers and consumers alike.

This does not mean that more conventional lines of economic analysis and enquiry are redundant or wrong. There are situations in which companies want to understand what consumers might be willing to pay for a specific product, or whether demand for that item might rise or fall a decade hence. Our point is that framing questions, responses and methods of analysis in these terms excludes others. In short, characterising markets in this way tells us nothing about either the present or the future world in which that product (or resource, or service) has value, not in general but as an outcome of specific practices, whether at home, at work or in moving around. Conceptualising demand in these terms takes us beyond the realm of regular economics, and of models and theories that revolve around provision and price. Amongst other things, such an approach forces us to recognise that social practices and the resources and appliances on which they depend are multiply interlinked. This is obvious when thinking about infrastructures that enable many practices at once, and when

thinking about how multiple practices are sequenced and scheduled through the day, over the week and across the year. In many European countries the evening peak in electricity demand is defined by the time when people cook their dinner and by what they do before and after that event. In France the peak is a bit later than it is in the UK because of different habits not only of eating, but also of watching TV and using the internet. The contention that the timing of demand is, in essence, an outcome of these kinds of societal rhythms has practical consequences. As well as helping to make sense of detailed variations in when consumption occurs, it generates new ways of defining flexibility: not as a property of an energy system, a person or a single practice, but as a feature of how complexes of practices interact at different temporal scales.

We discussed the material embedding of demand in Chapter 3, and considered matters of temporality in Chapter 4. Both are important in their own right, but when read together, these chapters take us into new territory. In essence they suggest that institutional and every day rhythms, from schooling and health care to eating and sleeping, are inscribed in the sizing and design of homes, cars and fridge freezers and in the capacities of electricity grids and of road and rail networks. By implication, the extent of demand and its timing should *not* be thought of as separate topics. In some ways this is obvious. The dimensions of freezers, car boots and kitchens represent forms of material capacity that are not arbitrary, but that relate to anticipated peaks in demand and to judgements about the scale, and also the rhythms of shopping and provisioning. In an article on 'social loading', Wilhite and Lutzenhiser (1999) write about how templates of imagined family life are reproduced in material form. In the example they give, cars are large enough to take extra passengers when the occasion arises. Likewise, ovens are sized to accommodate the Christmas turkey, even though that is cooked only once a year (Morley and Shove, 2014). These are relevant observations, but they are also just a start. Understanding how material and temporal arrangements shape each other has yet to become a central theme in social theory, and for the time being, questions like these are simply not on the radar in energy research and policy. With luck this book will help push these lines of enquiry up the agenda.

More straightforwardly, and also more immediately, all chapters suppose that demand is made and not simply met. This is a consistent and coherent position, but it is one that is at odds with conventional methods of treating demand as if it had an independent existence of its own. In writing about how present and future 'needs' are

estimated, especially in the transport sector, we have made much of the fact that the forms of analysis and forecasting involved generate specific pictures or visions of the future and that these have an active role in fostering the emergence of some but not other patterns of demand.

This works in different ways. To give just one example, the Department for Transport, advises that traffic surveys to support models are undertaken 'during a "neutral", or representative, month avoiding main and local holiday periods, local school holidays and half terms, and other abnormal traffic periods' (DfT, 2014: 11). Once these complicating factors have been excluded it turns out that there are only 80 neutral days in the year. The decision to focus on these is, at the same time, a decision to disregard swings and variations in what it is that people actually do. This is important in that investments that are made on the basis of analysing so-called neutral days are real in their effects. This would also be the case if different methods were used, and if the aim was to capture rather than obscure variation in demand. In other words, the point is not that there is, or could be some better method of knowing real needs. Rather, our claim is that since all such methods are implicated in constituting future demands, there should be greater transparency and more open discussion about what kinds of needs are envisaged, how, and by whom.

In taking discussions of demand in this direction we expand the scope for policy intervention. As we have seen, there is a tendency to take present patterns of energy and mobility demand for granted: to assume that current standards of comfort, or journey types will persist unchanged, and to conclude that the only option is to find more efficient ways of meeting these needs. Although not intended this way, the actions and strategies that follow are likely to sustain practices, standards and expectations that simply cannot be met in a low carbon future.

This is not the only way forward and as we have argued policy makers' capacities to shape demand could and should be mobilised in other ways. The next few paragraphs identify and illustrate types of policy intervention that would be consistent with our core propositions. Rather than introducing new cases and examples, we show what such an approach might entail in relation to food provisioning and the cold chain, online shopping and the four-day week.

Around 20% of UK domestic electricity demand is directly tied to the ways in which food is bought, sold and stored and to what foods are consumed and cooked. In this sector, policy responses have focused on the energy efficiency of the appliances involved, including

fridges and freezers along with microwaves and ovens. This approach has undoubtedly had some success: in the UK, in 2013, cold appliances were about twice as efficient as those sold in 2000 (Department of Energy and Climate Change, 2014). However, the volume of frozen and refrigerated space continues to increase in the UK and around the world. Twilley gives a sense of the scale and rate of change. In her words:

> China had 250 million cubic feet of refrigerated storage capacity in 2007; by 2017, the country is on track to have 20 times that. At five billion cubic feet, China will surpass even the United States, which has led the world in cold storage ever since artificial refrigeration was invented. And even that translates to only 3.7 cubic feet of cold storage per capita, or roughly a third of what Americans currently have.
>
> (Twilley, 2014)

The normalisation of chilled and frozen food means more refrigerated factory rooms; fridge and freezer cabinets in shops (supermarkets and cafés), and more and larger capacity for cold storage at home and on the move as food is transferred between these sites. So what kinds of policy interventions might stem this escalation of demand? Smart fridges and freezers that turn themselves off for parts of the day? A more efficient supply chain that reduces the time food spends in transit? However appealing strategies like these might be, they represent the next iteration of ideas that are already well established, and that have actually been part of the ratcheting up of demand. Crucially, they do not require any substantial change either to systems of provision or to related patterns of consumption.

A less fashionable but perhaps more effective strategy would be to consider food provision in the round. For example, a greater focus on localism might mean that fresh produce could be kept safely, but without the need for extensive networks of cold storage. Less reliance on meat and dairy might mean greater dependence on pulses and other foods that do not need to be kept cool. Steps might be taken to actively promote techniques of drying and preserving along with modified methods of food production. The need to keep so-called 'preserves' in a refrigerator once opened is a novel phenomenon: as we know, jam does not have to be made this way. By implication, effective policy intervention depends on reconfiguring complex webs of interaction in food provisioning, preparation, storage and diet. This might seem like an impossible task, but it is only very recently that

food systems have become so dependent on refrigeration. Since the nature of what people eat has changed rapidly over the last few decades and since it may change again in the coming years there is no reason to suppose that the future is fixed. This argues for policy approaches that challenge, rather than reinforce food systems and diets that require cooling.

Our second example concerns the relation between physical and online shopping and between in-store, home and third place delivery. In returning to this topic, we point to some of the routes through which non-energy policies shape patterns of mobility and energy demand and to the opportunities this presents.

There is evidence of a reduction in the number of trips people make to buy at least some sorts of items in store, but this has been partially offset by an increase in van traffic associated with home delivery. Since there is no Department of Shopping, and no 'home shopping policy' it might seem that trends like these are simply expressions of market forces, and as such beyond the realm of legitimate policy influence. On the other hand, methods of provision, delivery and consumption do not develop aside from existing and planned infrastructures, or from histories of design and urbanisation, all of which matter for the emergence of new forms. In recent years there has been a massive shift in the warehousing, ordering and dispatch function, led by companies like Amazon which seek to grab market share by offering subscription services with free next day, and increasingly same day, delivery. Their ambition of delivering to the door, and of doing so at a moment's notice is seemingly obvious, but this last 'mile' is exceptionally energy and mobility intensive, partly because the receiving end of the system is characterised by a hotchpotch of infrastructures including reception desks, log sheds, bins, rabbit hutches and letter boxes of different sizes.

Rather than standing back and letting online shopping develop around a single model of home delivery, policy makers could intervene with the aim of promoting other less demanding solutions. This might involve facilitating the development of consolidation points (e.g. shops and pubs, bus stations, libraries etc.) either directly or through favourable tax arrangements. Some of this is already happening with pick-up stores at old ticket offices in London Underground stations and in some chains of shops. However, more could be done to embed different forms of dropping off, collection and delivery. Compared with the fragmented patterns that are emerging today, previous methods, including mobile shops and other modes of local delivery, were organised around principles of route planning,

consolidation and off-peak timing. At the moment, deliveries of personal goods take place in busy city centres, at peak times and in already polluted areas. How fast do deliveries need to be from click to arrival? The shorter the lead times the more difficult the consolidation task. These are negotiable societal trade-offs as well as matters of operational research and optimisation. A much more coordinated approach to the spatial and temporal aspects of home delivery could alleviate rather than exacerbate these pressures.

Although it is possible to point to these and other opportunities, it is important to recognise that interventions along these lines represent efforts to modulate processes that already are in flux. As such, the consequences are impossible to predict with any certainty. However, the fact that the relation between online and bricks-and-mortar shopping continues to evolve is not a justification for inaction. In failing to engage with the energy and mobility implications of home delivery, policy makers have enabled online shopping to develop in ways that are more resource intensive than might have otherwise been the case.

Our final example comes from outside of the field of energy demand reduction and, as such, demonstrates the importance of invisible energy policy as described in Chapter 5. The history of the working week is complicated and varied (Zerubavel, 1985), but the pattern of a two-day weekend is now well established. However, other arrangements are possible, including that of a four-day or 32-hour week (Labour Party, 2019). What difference might a reduction in working hours make to energy and mobility demand? Building on the ideas introduced in Chapter 4, much depends on how such a policy is implemented. For example, would it apply to all sectors of society, or only to some? Would working patterns be reorganised so that current levels of service provision are maintained, reduced or increased? Would a reduction in time at work be achieved through the creation of a three-day weekend, and if so would that be Friday to Sunday, or Saturday to Monday, or would it be staggered across the week?

These details are hugely important for how activities are sequenced and scheduled and for the infrastructures on which they depend. For example, a three-day weekend would not reduce peak hour traffic on working days, but a more differentiated pattern might have this effect. More varied configurations of working hours across a five-day week might diminish the spikes in energy demand that occur when many people return home and prepare dinner at the same time in the evening. Of course, a reduction in societal synchronisation might also limit the scope for reducing peak hour traffic by car sharing. Either way, it would be important to consider the seasonal effects of reduced working hours.

Since activities are already more condensed in winter than summer due to daylight hours and temperatures, a four-day week is likely to have different consequences at different times of year. In posing these questions we are not making the case for or against such a policy, but we are suggesting that anyone thinking about the implications of a four-day week or a seven-day health service or a 24/7 supermarket needs to think about how such interventions might reshape a very wide range of societal rhythms. This is not just a matter of averting unintended consequences. More positively, the point is that there are opportunities to design and steer policies in ways that reduce rather than increase energy demand.

That said, there are no simple policy levers to pull. Demand making is clearly not the province of one policy area, or of policy alone. Some might take this to be a dispiriting conclusion. We argue otherwise. Recognising that policy intervention is risky and complex is not an argument for standing back or for accepting, and thus perpetuating the status quo. Not having an explicit demand policy is in fact a demand policy, just an unthinking one. Looking ahead, the challenge is to understand how different kinds of demand emerge, persist and disappear, and to identify when, where and how policy makers might intervene. Shaping and fostering social practices that are systematically less energy and mobility demanding than those that exist today is likely to involve new forms of collaboration across policy areas. It also requires an appreciation both of the histories of contemporary practices, and of the openness of the future.

This is not the case at present, or at least it is not for those forms of policy making that take current standards of living for granted, or that are informed by abstract models that obscure, and that are blind to more fundamental changes in demand (see Chapter 5). In less than a generation, interpretations of what it means to be comfortable, of what counts as a normal diet or of the sorts of objects that people might make or use or own have changed spectacularly. It would be very odd indeed if these ongoing transformations were to come to a sudden halt, or if the trends of the last five decades were to continue along the same path. This does not mean that policy makers should try to predict the future. The more important point is to recognise that future trajectories are being made today, and that whatever they do, policy makers are part of that process. As represented here, new patterns of demand cannot be brought about by act of will, but neither are they simply outcomes of external forces. Instead interventions by policy makers and others form part of ongoing and inherently dynamic processes that are also shaped by the historical layering of previous policies, infrastructures and material arrangements.

Throughout this book we have concentrated on examples that relate to energy and transport on the grounds that engendering societal changes that promise to reduce, rather than increase carbon emissions is a current and pressing challenge. However, our approach to defining and analysing the foundations of demand, and the questions we raise are of wider significance. As such, much of what we have had to say is important and relevant for organisations and institutions operating in different sectors (for example, health, education and other public services), and at different scales. We are convinced that the arguments we have made and the propositions we have explored are coherent and compelling and we hope that they are also useful and inspiring. Even so, it would be astonishing if the ideas we have set out were to be taken up and adopted wholesale. As we have noticed, our conceptualisation of demand runs against the grain of currently dominant paradigms. In addition, many of our suggestions suppose and may even require political and institutional practices that do not yet exist. At the same time, it is increasingly obvious that if there is to be any hope of moving towards a low carbon society the questions we address – how is demand made, how does it change and how might it be steered? – have to take centre stage.

References

Adam, B. (1990). *Time and Social Theory*. Cambridge: Polity Press.

Ajzen, I. (1991). The theory of planned behavior. *Organizational Behavior and Human Decision Processes*, 50(2), 179–211.

Akrich, M. (1992). The de-scription of technical objects. In W. Bijker & J. Law (eds.) *Shaping Technology, Building Society*. Cambridge, MA: MIT Press. 205–225.

Allen, J., Piecyk, M., & Piotrowska, M. (2017). Analysis of online shopping and home delivery in the UK, London, FTC2050 Project Report, www.ftc2050. com/reports/Online_shopping_and_home_delivery_in_the_UK_final_version_Feb_2017.pdf.

Anderson, B. (2016). Laundry, energy and time: insights from 20 years of time-use diary data in the United Kingdom. *Energy Research & Social Science*, 22, 125–136.

Appadurai, A. (1994). Commodities and the politics of value. In S. M. Pearce (ed.) *Interpreting Objects and Collections*, 76–91.

Appleby, J. (2013) Spending on health and social care over the next 50 years: why think long term? *The Kings Fund*. Available at: www.kingsfund.org.uk/sites/default/files/field/field_publication_file/Spending%20on%20health%20..%2050%20years%20low%20res%20for%20web.pdf, Accessed 10 September 2019.

Arnould, E. & Thompson, C. J. (2005). Consumer culture theory (CCT): twenty years of research. *Journal of Consumer Research*, 31(4), 868–882.

Asif, M., & Muneer, T. (2007). Energy supply, its demand and security issues for developed and emerging economies. *Renewable and Sustainable Energy Reviews*, 11 (7), 1388–1413.

Association for Decentralised Energy. (2016). *Flexibility on demand: giving customers control to secure our electricity system*. Available at: www.theade.co.uk/assets/docs/resources/Flexibility_on_demand_full_report.pdf, Accessed 19 September 2019.

Aucoin, P. (1990). Administrative reform in public management: paradigms, principles, paradoxes and pendulums. *Governance*, 3(2), 115–137.

Bache, I., & Flinders, M. (2004). *Multi-Level Governance*. Oxford: Oxford University Press.

Barr, S. (2006). Environmental action in the home: investigating the 'value-action' gap. *Geography*, 91(1), 43–54.

Barr, S., & Prillwitz, J. (2014). A smarter choice? Exploring the behaviour change agenda for environmentally sustainable mobility. *Environment and Planning Part C*, 32(1), 1–19.

Bastian, A., Börjesson, M. & Eliasson, J. (2016). Explaining 'peak car' with economic variables. *Transportation Research Part A*, C(88), 236–250.

Baudrillard, J. (2016). *The Consumer Society: Myths and Structures*. London: Sage.

BEIS. (2019). *Updated energy and emissions projections 2018*. Department for Business, Energy and Industrial Strategy. Available at: www.gov.uk/government/publications/updated-energy-and-emissions-projections-2018, Accessed 10 September 2019.

Belk, R. (1988). Possessions and the extended self. *Journal of Consumer Research*, 15(2), 139–168.

Birkland, T. (2016). *An Introduction to the Policy Process: Theories, Concepts and Models of Public Policy Making*, Fourth Edition. London: Routledge.

Blue, S. (2017). Institutional rhythms: combining practice theory and rhythmanalysis to conceptualise processes of institutionalisation. *Time & Society*, 28(3), 922–950.

Blue, S. (2018). Reducing demand for energy in hospitals: opportunities for and limits to temporal coordination. In A. Hui, R. Day & G. Walker (eds.) *Demanding Energy*. Basingstoke: Palgrave Macmillan. 313–337.

Boswell, C., Geddes, A. & Scholten, P. (2011). The role of narratives in migration policy-making: a research framework. *British Journal of Politics and International Relations*, 13(1), 1–11.

Bourdieu, P. (1984). *Distinction: A Social Critique of the Judgement of Taste*. Cambridge, MA: Harvard University Press.

Brundtland, G. H., Khalid, M., Agnelli, S., Al-Athel, S. & Chidzero, B. (1987). *Our Common Future*. New York: United Nations.

Büchs, M., & Koch, M. (2019). Challenges for the degrowth transition: the debate about wellbeing. *Futures*, 105, 155–165.

Butler, C., Parkhill, K. & Luzecka, P. (2018). Rethinking energy demand governance: exploring impact beyond 'energy' policy. *Energy Research and Social Science*, 36, 70–78.

Callon, M. (1984). Some elements of a sociology of translation: domestication of the scallops and the fishermen of St Brieuc Bay. *The Sociological Review*, 32(1), 196–233.

Callon, M. (1998). *The Laws of the Markets*. Hoboken, NJ: Blackwell Publishers.

Callon, M., & Muniesa, F. (2005). Peripheral vision: economic markets as calculative collective devices. *Organization Studies*, 26(8), 1229–1250.

Callon, M., Millo, Y. & Muniesa, F. (2007). *Market Devices* (No. halshs-00177891).

Carbon Trust. (2016). *An analysis of electricity system flexibility for Great Britain*. Imperial College London. Available at: https://assets.publishing.service.gov.uk/government/uploads/system/uploads/attachment_data/file/568982/An_analysis_of_electricity_flexibility_for_Great_Britain.pdf, Accessed 19 September 2019.

Carmichael, R. (2019). Behaviour change, public engagement and net zero. Report for the Committee on Climate Change. Available at: www.theccc.org.uk/wp-

content/uploads/2019/10/Behaviour-change-public-engagement-and-Net-Zero-Imperial-College-London.pdf, Accessed 9 January 2020.

Castells, M. (1996). *The Information Age* (Vol. 98). Oxford: Blackwell Publishers.

CCC. (2018). *Reducing UK Emissions – 2018 Progress Report to Parliament*. Committee on Climate Change. Available at: www.theccc.org.uk/publication/reducing-uk-emissions-2018-progress-report-to-parliament/, Accessed 10 September 2019.

Chappells, H., & Trentmann, F. (2018). Disruption in and across time. In E. Shove & F. Trentmann (eds.) *Infrastructures in Practice: The Dynamics of Demand in Networked Societies*. London: Routledge. 197–210.

Chatterjee, K., Goodwin, P., Schwanen, T., Clark, B., Jain, J., Melia, S., Middleton, J., Plyushteva, A., Ricci, M., Santos, G. & Stokes, G. (2018). *Young People's Travel – What's changed and why? Review and analysis*. Report to Department for Transport. Bristol: UWE. Available at: www.gov.uk/government/publications/young-peoples-travel-whats-changed-and-why, Accessed 10 September 2019.

Christensen, T., & Lægreid, P. (2007). The whole of government approach to public sector reform. *Public Administration Review*, 67(6), 1059–1066.

Cochoy, F., Hagberg, J. & Canu, R. (2015). The forgotten role of pedestrian transportation in urban life: insights from a visual comparative archaeology (Gothenburg and Toulouse, 1875–2011). *Urban Studies*, 52(12), 2267–2286.

Committee on Climate Change. (2019). *Net Zero. The UK's contribution to stopping global warming*. Available at: www.theccc.org.uk/publication/net-zero-the-uks-contribution-to-stopping-global-warming/, Accessed 2 July 2019.

Cooper, G. (2002). *Air-conditioning America: Engineers and the Controlled Environment, 1900–1960*. Baltimore, MD: JHU Press.

Coutard, O., & Shove, E. (2018). Infrastructures, practices and the dynamics of demand. In E. Shove & F. Trentmann (eds.) *Infrastructures in Practice: The Dynamics of Demand in Networked Societies*. London: Routledge. 10–23.

Cox, E., Royston, S. & Selby, J. (2016). *The impacts of non-energy policies on the energy system: a scoping paper*. UKERC. Available at: www.ukerc.ac.uk/publications/impact-of-non-energy-policies-on-energy-systems.html, Accessed 10 September 2019.

Darnton, A., Verplanken, B., White, P. & Whitmarsh, L. (2011). *Habits, Routines and Sustainable Lifestyles, A Summary Report to the Department for Environment, Food and Rural Affairs*. London: AD Research & Analysis for Defra.

David, P. (1985). Clio and the Economics of QWERTY. *The American Economic Review*, 75(2), 332–337.

Day, R., Walker, G. & Simcock, N. (2016). Conceptualising energy use and energy poverty using a capabilities framework. *Energy Policy*, 93, 255–264.

De Decker, K. (2009). Wind powered factories: history (and future) of industrial windmills. *Low-Tech Magazine*. Available at: https://solar.lowtechmagazine.com/2009/10/history-of-industrial-windmills.html, Accessed 19 September 2019.

de La Bruhèze, A. & van Otterloo, A. (2004). The Milky way: infrastructures and the shaping of milk chains. *History and Technology*, 20(3), 249–269.

Defra. (2017). *Greening Government Commitments 2015–2016 Annual Report*. Available at: www.gov.uk/government/publications/greening-government-commitments-2015-to-2016-annual-report, Accessed 19 September 2019.

Department of Energy and Climate Change. (2014). Energy efficient products – helping us cut energy use. Available at: www.gov.uk/government/publications/energy-efficient-products-helping-us-cut-energy-use, Accessed 9 January 2020.

DfT. (2014). *TAG Unit M1.2*. London: Department for Transport. Available at: https://assets.publishing.service.gov.uk/government/uploads/system/uploads/attachment_data/file/427119/webtag-tag-unit-m1-2-data-sources-and-surveys.pdf, Accessed 10 September 2019.

DfT. (2018a). *Car Travel Econometrics. Moving Britain Ahead*. London: Department for Transport. Available at: www.gov.uk/government/publications/car-travel-econometrics, Accessed 19 September 2019.

DfT. (2018b). *Road Traffic Forecasts 2018*. London: Department for Transport. Available at: www.gov.uk/government/publications/road-traffic-forecasts-2018, Accessed 10 September 2019.

Di Giulio, A., & Fuchs, D. (2014). Sustainable consumption corridors: concept, objections, and responses. *GAIA-Ecological Perspectives for Science and Society*, 23(3), 184–192.

Docherty, I., Shaw, J., Marsden, G. & Anable, J. (2018). The curious death – and life? – of British transport policy. *Environment and Planning C: Government and Policy*, 36(8), 1458–1479.

Doyal, L., & Gough, I. (1991). *A Theory of Human Need*. Basingstoke: The Macmillan Press Ltd.

Drysdale, B., Wu, J. & Jenkins, N. (2015). Flexible demand in the GB domestic electricity sector in 2030. *Applied Energy*, 139, 281–290.

Dupuis, M. (2002). *Nature's Perfect Food: How Milk Became America's Drink*. New York: NYU Press.

Durkheim, E., & Lukes, S. (2013). *Durkheim: The Division of Labour in Society*. Basingstoke: Palgrave Macmillan.

Dye, T. (2012). *Understanding Public Policy*, Fourteenth Edition. London: Pearson.

EA (2017). *Drought Response: Our Framework for England*. Environment Agency. Available at: https://assets.publishing.service.gov.uk/government/uploads/system/uploads/attachment_data/file/625006/LIT_10104.pdf, Accessed 10 September 2019.

EIA. (2016). *International Energy Outlook 2016*. U.S. Energy Information Administration. Chapter 8. Available at: www.eia.gov/outlooks/ieo/pdf/transportation.pdf, Accessed 19 September 2019.

Ellegård, K., & Svedin, U. (2012). Torsten Hägerstrand's time-geography as the cradle of the activity approach in transport geography. *Journal of Transport Geography*, 23, 17–25.

Faulconbridge, J., Cass, N. & Connaughton, J. (2017). How market standards affect building design: the case of low energy design in commercial offices. *Environment and Planning A*, 50(3), 627–650.

Featherstone, M. (2007). *Consumer Culture and Postmodernism*. London: Sage.

Forman, P. (2017). *Securing Natural Gas: Entity-attentive Security Research*. PhD Thesis. Durham: Durham University.

Forty, A. (1986). *Objects of Desire: Design and Society, 1750–1980*. London: Thames and Hudson.

Fouquet, R. (2014). Long-run demand for energy services: income and price elasticities over two hundred years. *Review of Environmental Economics and Policy*, 8(2), 186–207.

Freidberg, S. (2015). Moral economies and the cold chain. *Historical Research*, 88 (239), 125–137.

Geels, F. & Schot, J. (2007). Typology of sociotechnical transition pathways. *Research Policy*, 36(3), 399–417.

Geertz, C. (1973). *The Interpretation of Cultures*. New York: Basic Books.

Gell, A. (1986). Newcomers to the world of goods: consumption among the Muria Gonds. In A. Appadurai (ed.) *The Social Life of Things: Commodities in Cultural Perspective*. Cambridge: Cambridge University Press. 110–138.

Geng, Y., Sarkis, J., Ulgiati, S. & Zhang, P. (2013). Measuring China's circular economy. *Science*, 339(6127), 1526–1527.

Giddens, A. (1979). *Central Problems in Social Theory: Action, Structure, and Contradiction in Social Analysis*. Berkeley, CA: University of California Press.

Giddens, A. (1984). *The Constitution of Society*. Cambridge: Polity Press.

Glennie, P., & Thrift, N. (2009). *Shaping the Day: A History of Timekeeping in England and Wales 1300–1800*. Oxford: Oxford University Press.

Godin, B., & Lane, J. (2013). Pushes and pulls: hi(S)tory of the demand-pull model of innovation. *Science, Technology, & Human Values*, 38(5), 621–654.

Goodwin, P., Hallett, S., Kenny, F. & Stokes, G. (1991) *Transport: The New Realism*. Report to the Rees Jeffreys Road Fund for discussion at the 'Transport – The New Realism' conference, Church House, London, 21st March 1991 TSU Ref: 624, Available at: www.tsu.ox.ac.uk/pubs/1062-goodwin-hallett-kenny-stokes.pdf, Accessed 10 September 2019.

Goodwin, P., & Lyons, G. (2009). Public attitudes to transport: interpreting the evidence. *Transportation Planning and Technology*, 33(1), 3–17.

Goulden, M., Ryley, T. & Dingwall, R. (2014). Beyond 'predict and provide': UK transport, the growth paradigm and climate change. *Transport Policy*, 32, 139–147.

Graham, S., & Marvin, S. (2001). *Splintering Urbanism: Networked Infrastructures, Technological Mobilities and the Urban Condition*. London: Routledge.

Gronow, J., & Warde, A. (eds.). (2001). *Ordinary Consumption* (vol. 2). Brighton: Psychology Press.

Hand, M., & Shove, E. (2007). Condensing practices: ways of living with a freezer. *Journal of Consumer Culture*, 7(1), 79–104.

Hand, M., Shove, E. & Southerton, D. (2005). Explaining showering: a discussion of the material, conventional, and temporal dimensions of practice. *Sociological Research Online*, 10(2), 1–13.

Haraway, D. (2006). A cyborg manifesto: science, technology, and socialist-feminism in the late 20th century. In J. Weiss, J. Nolan, J. Hunsinger & P. Trifonas (eds.) *The International Handbook of Virtual Learning Environments*. Dordrecht: Springer. 117–158.

Hardt, L., Barrett, J., Brockway, P., Foxon, T., Heun, M. K., Owen, A. & Taylor, P. (2017). Outsourcing or efficiency? Investigating the decline in final energy consumption in the UK productive sectors. *Energy Procedia* 142, 2409–2414.

Heikkurinen, P. (2018). Degrowth by means of technology? A treatise for an ethos of releasement. *Journal of Cleaner Production*, 197, 1654–1665.

Heiskanen, E., & Pantzar, M. (1997). Toward sustainable consumption: two new perspectives. *Journal of Consumer Policy*, 20(4), 409–442.

Hills, P. J. (1996). What is induced traffic? *Transportation*, 23, 5–16.

Hitchings, R. (2010). Seasonal climate change and the indoor city worker. *Transactions of the Institute of British Geographers*, 35(2), 282–298.

HM Government. (2018). *Upgrading Our Energy System: Smart Systems and Flexibility Plan: Progress Update*. Available at: https://assets.publishing.service.gov.uk/govern ment/uploads/system/uploads/attachment_data/file/756051/ssfp-progress-update.pdf, Accessed 19 September 2019.

HM Treasury. (2011). *The National Infrastructure Plan 2011*. London: HM Treasury. Available at: https://assets.publishing.service.gov.uk/government/uploads/system/uploads/attachment_data/file/188337/nip_2011.pdf, Accessed 10 September 2019.

Hobson, K. (2003). Thinking habits into action: the role of knowledge and process in questioning household consumption practices. *Local Environment*, 8(1), 95–112.

Holbrook, M. & Hirschman, E. (1982). The experiential aspects of consumption: consumer fantasies, feelings, and fun. *Journal of Consumer Research*, 9(2), 132–140.

Holt, D. (1995). How consumers consume: a typology of consumption practices. *Journal of Consumer Research*, 22(1), 1–16.

Holt, D. (1997). Poststructuralist lifestyle analysis: conceptualizing the social patterning of consumption in postmodernity. *Journal of Consumer Research*, 23(4), 326–350.

Holzer, B. (2006). Political consumerism between individual choice and collective action: social movements, role mobilization and signalling. *International Journal of Consumer Studies*, 30(5), 405–415.

Howlett, M., Ramesh, M. & Perl, A. (2009). *Studying Public Policy: Policy Cycles and Policy Subsystems*, Third Edition. Oxford: Oxford University Press.

Hughes, T. (1993). *Networks of Power: Electrification in Western Society, 1880–1930*. Baltimore, MD: JHU Press.

Hui, A., Schatzki, T. & Shove, E. (2017). *The Nexus of Practices: Connections, Constellations, Practitioners*. London: Routledge.

IEA. (2018). *The world energy outlook 2018*. International energy agency. Available at: www.iea.org/weo, Accessed 19 September 2019.

Ingold, T. (2011). *Being Alive: Essays on Movement, Knowledge and Description*. London: Routledge.

Ingold, T. (2012). Toward an ecology of materials. *Annual Review of Anthropology*, 41, 427–442.

IPA. (2018). *Annual report on major projects 2017–2018*. London: Infrastructure and Projects Authority. Available at: www.gov.uk/government/publications/infrastructure-and-projects-authority-annual-report-2018, Accessed 10 September 2019.

Jenkins, K., & Hobson, D. (eds.). (2018). *Transitions in Energy Efficiency and Demand – The Emergence, Diffusion and Impact of Low-Carbon Innovation*. London: Routledge.

Kaufmann, J.-C. (1998). *Dirty Linen: Couples and Their Laundry*. London: Middlesex University Press.

Kemmis, S. (2009). What is professional practice? Recognising and respecting diversity in understandings of practice. In C. Kanes (ed.) *Elaborating Professionalism*. Dordrecht: Springer. 139–165.

Kuijer, L., & De Jong, A. (2012). Identifying design opportunities for reduced household resource consumption: exploring practices of thermal comfort. *Journal of Design Research*, 14, 10(1–2), 67–85.

Kuijer, L., & Watson, M. (2017). 'That's when we started using the living room': lessons from a local history of domestic heating in the United Kingdom. *Energy Research & Social Science*, 28, 77–85.

Labanca, N. (ed.). 2017. *Complex Systems and Social Practices in Energy Transitions: Framing Energy Sustainability in the Time of Renewables*. London: Springer.

Labour Party. (2019). *It's time for real change. The Labour Party manifesto 2019*. Available at: https://labour.org.uk/wp-content/uploads/2019/11/Real-Change-Labour-Manifesto-2019.pdf, Accessed 9 January 2020.

Latour, B. (1992). Where are the missing masses? The sociology of a few mundane artifacts. In W. Bijker & J. Law (eds.) *Shaping Technology-Building Society. Studies in Sociotechnical Change*. Cambridge, MA: MIT Press. 225–259.

Latour, B. (2005). *Reassembling the Social: An Introduction to Actor-Network-Theory*. Oxford, New York: Oxford University Press.

Le Vine, S., Polak, J. & Humphrey, A. (2017). *Commuting Trends in England 1988–2015*. Department for Transport. Available at: https://assets.publishing.service.gov.uk/government/uploads/system/uploads/attachment_data/file/657839/commuting-in-england-1988-2015.pdf, Accessed 15 May 2019.

Lefebvre, H. (2004). *Rhythmanalysis: Space, Time, and Everyday Life*. London: Continuum.

Lipsky, M. (2010). *Street-Level Bureaucracy: Dilemmas of the Individual in Public Services*, 30th Anniversary Edition. New York: Russell Sage Foundation.

MacIntyre, A. (2013). *After Virtue*. London: A&C Black.

Mackay, I., Batten, R. & Franz, I. (2003). Garages & Filling Stations, www.igg.org.uk/rail/00-app1/gar-pet.htm.

Maller, C., & Strengers, C. (eds.). 2018. *Social Practices and Dynamic Non-humans: Nature, Materials and Technologies*. Cham: Palgrave Macmillan.

Marsden, G. (2019). Rebound. In J. Rinkinen, E. Shove & J. Torriti (eds.) *Energy Fables: Challenging Ideas in the Energy Sector*. London: Routledge. 39–47.

Marsden, G., Dales, J., Jones, P., Seagriff, E. & Spurling, N. (2018). All change? The future of travel demand and its implications for policy and planning, First

Report of the Commission on Travel Demand, Available at: www.demand.ac. uk/commission-on-travel-demand/, Accessed 10 September 2019.

Marsden, G., & McDonald, N. (2019). Institutional issues in planning for more uncertain futures. *Transportation*, 46(4), 1075–1092.

Marsden, G., Mullen, C. A., Bache, I. et al. (2014). Carbon reduction and travel behaviour: discourses, disputes and contradictions in governance. *Transport Policy*, 35, 71–78.

Mattioli, G., Anable, J. & Vrotsou, K. (2016). Car dependent practices: findings from a sequence pattern mining study of UK time use data. *Transportation Research Part A: Policy and Practice*, 89, 56–72.

McConnell, C., Brue, S. & Flynn, S. (2009). *Economics: Principles, Problems, and Policies*. Irwin: Boston McGraw-Hill.

McCracken, G. (1990). *Culture and Consumption: New Approaches to the Symbolic Character of Consumer Goods and Activities*. Bloomington, IN: Indiana University Press.

McFarlane, C. (2011). The city as assemblage: dwelling and urban space. *Environment and Planning D: Society and Space*, 29(4), 649–671.

Meadows, D., Meadows, D., Randers, J. & Behrens, W. (1972). *The Limits to Growth: A Report to the Club of Rome*. Washington, DC: Potomac Associates.

Meier, A., & Siderius, H. (2017). Should the next standby power target be 0-watt? *Lawrence Berkeley National Laboratory*. Available at: https://escholarship.org/uc/ item/566951pn, Accessed 19 September 2019.

Miller, D. (2002). Turning Callon the right way up. *Economy and Society*, 31(2), 218–233.

Mokhtarian, P. & Salomon, I. (2001). How derived is the demand for travel? Some conceptual and measurement considerations. *Transportation Research Part A: Policy and Practice*, 35(8), 695–719.

Monstadt, J. (2009). Conceptualizing the political ecology of urban infrastructures: insights from technology and urban studies. *Environment and Planning A*, 41(8), 1924–1942.

Morley, J. (2017). Technologies within and beyond practices. In A. Hui, T. Schatzki & E. Shove (eds.) *The Nexus of Practices: Connections, Constellations, Practitioners*. London: Routledge. 81–97.

Morley, J. (2018). Rethinking energy services: the concept of 'meta-service' and implications for demand reduction and servicizing policy. *Energy Policy*, 122, 563–569.

Morley, J. (2019). Energy services. In J. Rinkinen, E. Shove & J. Torriti (eds.) *Energy Fables: Challenging Ideas in the Energy Sector*. London: Routledge. 15–26.

Morley, J., & Shove, E. (2014). Size is everything at Christmas and your oven is no exception. *The Conversation*. Available at: http://theconversation.com/size-is-everything-at-christmas-and-your-oven-is-no-exception-35689, Accessed 15 May 2019.

Mullen, C., & Marsden, G. (2015). Transport, economic competitiveness and competition: a city perspective. *Journal of Transport Geography*, 49, 1–8.

National Grid (2017). *Power responsive: demand side flexibility annual report 2017.* Available at: http://powerresponsive.com/wp-content/uploads/2018/02/Power-Responsive-Annual-Report-2017.pdf, Accessed 19 September 2019.

Nye, D. E. (2010). *When the Lights Went Out: A History of Blackouts in America.* Cambridge, MA: MIT Press.

Ofcom. (2018). *Communications market report 2018.* Available at: www.ofcom.org.uk/research-and-data/multi-sector-research/cmr/cmr-2018/interactive, Accessed 19 September 2019.

Owens, S. (1995). From 'predict and provide' to 'predict and prevent'?: pricing and planning in transport policy. *Transport Policy,* 2(1), 43–49.

Pollitt, C. (2008). *Time, Policy, Management: Governing with the Past.* Oxford: Oxford University Press.

Powells, G., Bulkeley, H., Bell, S. & Judson, E. (2014). Peak electricity demand and the flexibility of everyday life. *Geoforum,* 55, 43–52.

Pred, A. (1981). Social reproduction and the time-geography of everyday life. *Geografiska Annaler. Series B. Human Geography,* 63(1), 5–22.

Putnam, R. (2001). *Bowling Alone: The Collapse and Revival of American Community.* New York: Simon & Schuster.

Rahman, S. (2011). *Temperature correction of energy statistics Office for National Statistics.* Available at: www.gov.uk/government/publications/temperature-correction-of-energy-statistics-from-the-office-of-national-statistics, Accessed 19 September 2019.

Reardon, L., & Marsden, G. (2020). Exploring the role of the state in the depoliticisation of UK transport policy. *Policy and Politics,* 48(2), 223–240. doi: 10.1332/030557319X15707904263616.

Reckwitz, A. (2002). Toward a theory of social practices: a development in culturalist theorizing. *European Journal of Social Theory,* 5(2), 243–263.

Rhodes, R. (2007). Understanding governance: ten years on. *Organization Studies,* 28(8), 1243–1264.

Rinkinen, J. (2015). *Demanding Energy in Everyday Life: Insights from Wood Heating into Theories of Social Practice.* PhD Thesis. Helsinki: Aalto School of Business.

Rinkinen, J. (2018). Chopping, stacking and burning wood: rhythms and variations in provision. In E. Shove & F. Trentmann (eds.) *Infrastructures in Practice.* London: Routledge. 48–57.

Rinkinen, J. & Jalas, M. (2017). Moving home: houses, new occupants and the formation of heating practices. *Building Research & Information,* 45(3), 293–302.

Rinkinen, J., Jalas, M. & Shove, E. (2015). Object relations in accounts of everyday life. *Sociology,* 49(5), 870–885.

Rinkinen, J., Shove, E. & Smits, M. (2017). Cold chains in Hanoi and Bangkok: changing systems of provision and practice. *Journal of Consumer Culture,* 19(3), 379–397.

Rittel, H. & Webber, M. (1973). Dilemmas in a general theory of planning. *Policy Sciences,* 4(2), 155–169.

Rochefort, D. & Cobb, R. (1993). Problem definition, agenda access and policy choice. *Policy Studies Journal,* 21(1), 56–71.

Royston, S., Selby, J. & Shove, E. (2018). Invisible energy policies: a new agenda for energy demand reduction. *Energy Policy*, 123, 127–135.

Saccani, N., Perona, M. & Bachetti, A. (2017). The total cost of ownership of durable consumer goods: a conceptual model and an empirical application. *International Journal of Production Economics*, 183(A), 1–13.

Saddler, H. (2013) What can we learn from looking at electricity use on Christmas day? *Conversation*. Available at: https://theconversation.com/what-can-we-learn-from-looking-at-electricity-use-on-christmas-day-20503, Accessed 20 September 2019.

Sassatelli, R. (2007). *Consumer Culture: History, Theory and Politics*. London: Sage.

Schatzki, T. (2010). Materiality and social life. *Nature and Culture*, 5(2), 123–149.

Schatzki, T. (2016). Keeping track of large phenomena. *Geographische Zeitschrift*, 104 (1), 4–24.

Schatzki, T. (2002). *The Site of the Social: A Philosophical Account of the Constitution of Social Life and Change*. Pennsylvania, PA: Penn State Press.

Schatzki, T. (2011). Where the action is (on large social phenomena such as sociotechnical regimes). *Sustainable Practices Research Group, Working Paper*, 1, 1–31.

Schön, D. & Rein, M. (1994). *Frame Reflection: Toward the Resolution of Intractable Policy Controversies*. New York: Basic Books.

Schot, J., Kanger, L. & Verbong, G. (2016). The roles of users in shaping transitions to new energy systems. *Nature Energy*, 1(5), 16054.

Sennett, R. (2011). *The Corrosion of Character: The Personal Consequences of Work in the New Capitalism*. New York: W. W. Norton.

Shove, E. (2003). *Converging Conventions of Comfort, Cleanliness and Convenience*. Oxford: Berg.

Shove, E. (2009). Everyday practice and the production and consumption of time. In E. Shove, F. Trentmann & R. Wilk (eds.) *Time, Consumption and Everyday Life: Practice, Materiality and Culture*. Oxford: Berg. 17–35.

Shove, E. (2010). Beyond the ABC: climate change policy and theories of social change. *Environment and Planning A*, 42(6), 1273–1285.

Shove, E. (2017). Matters of practice. In A. Hui, T. Schatzki & E. Shove (eds.) *The Nexus of Practices: Connections, Constellations, Practitioners*. London: Routledge. 155–168.

Shove, E. (2018). What is wrong with energy efficiency? *Building Research & Information*, 46(7), 779–789.

Shove, E., Chappells, H. & Lutzenhiser, L. (2009a). *Comfort in a Lower Carbon Society*. London: Routledge.

Shove, E. & Pantzar, M. (2005). Consumers, producers and practices: understanding the invention and reinvention of Nordic walking. *Journal of Consumer Culture*, 5 (1), 43–64.

Shove, E., Pantzar, M. & Watson, M. (2012). *The Dynamics of Social Practice: Everyday Life and How It Changes*. London: Sage.

Shove, E. & Southerton, D. (2000). Defrosting the freezer: from novelty to convenience: a narrative of normalization. *Journal of Material Culture*, 5(3), 301–319.

Shove, E. & Trentmann, F. (eds.). (2018). *Infrastructures in Practice: The Dynamics of Demand in Networked Societies*. London: Routledge.

Shove, E., Trentmann, F. & Wilk, R. (2009b). *Time, Consumption and Everyday Life: Practice, Materiality and Culture*. Oxford: Berg.

Shove, E. & Walker, G. (2014). What is energy for? Social practice and energy demand. *Theory, Culture & Society*, 31(5), 41–58.

Shove, E. & Warde, A. (2002). Inconspicuous consumption: the sociology of consumption, lifestyles and the environment. *Sociological Theory and the Environment: Classical Foundations, Contemporary Insights*, 230, 51.

Shove, E. Watson, M., Hand, M. & Ingram, J. (2007). *The Design of Everyday Life*. Oxford: Berg.

Shove, E., Watson, M. & Spurling, N. (2015). Conceptualizing connections: energy demand, infrastructures and social practices. *European Journal of Social Theory*, 18(3), 274–287.

Silvast, A. (2018). Co-constituting supply and demand: managing electricity in two neighbouring control rooms. In E. Shove & F. Trentmann (eds.) *Infrastructures in Practice: The Dynamics of Demand in Networked Societies*. London: Routledge. 171–183.

Silverstone, R. & Hirsch, E. (eds.). (1992). *Consuming Technologies: Media and Information in Domestic Spaces*. London: Routledge.

Singleton, A., Dolega, L., Riddlesden, D. & Longley, P. (2016). Measuring the spatial vulnerability of retail centres to online consumption through a framework of e-resilience. *Geoforum*, 69, 5–18.

Slater, D. (2002). Capturing markets from the economists. In P. du Gay & M. Pryke (eds.) *Cultural Economy*. London: Sage. 59–77.

Smale, R., van Vliet, B. & Spaargaren, G. (2017). When social practices meet smart grids: flexibility, grid management, and domestic consumption in The Netherlands. *Energy Research & Social Science*, 34, 132–140.

Smith, A. (1977). *The Wealth of Nations*. London: Dent.

Sorrell, S. (2007). *The Rebound Effect: An Assessment of the Evidence for Economy-Wide Energy Savings from Improved Energy Efficiency*. A report produced by the Sussex Energy Group. Available at: https://pdfs.semanticscholar.org/8e52/60a35163402b6ada126baddc023966252618.pdf, Accessed 20 September 2019.

Southerton, D. (2003). 'Squeezing time' – allocating practices, coordinating networks and scheduling society. *Time & Society*, 12(1), 5–25.

Southerton, D. (2009). Re-ordering temporal rhythms. Coordinating daily practices in the UK in 1937 and 2000. In E. Shove, F. Trentmann & R. Wilk (eds.) *Time, Consumption and Everyday Life: Practice, Materiality and Culture*. Oxford: Berg. 49–63.

Spaargaren, G. (2011). Theories of practices: agency, technology, and culture: exploring the relevance of practice theories for the governance of sustainable consumption practices in the new world-order. *Global Environmental Change*, 21(3), 813–822.

Spurling, N. (2018). Matters of time: materiality and the changing temporal organisation of everyday energy consumption. *Journal of Consumer Culture*, May 4, 2018, online.

Star, S. (1999). The ethnography of infrastructure. *American Behavioral Scientist*, 43 (3), 377–391.

Stead, D. (2008). Institutional aspects of integrating transport, environment and health policies. *Transport Policy*, 15(3), 139–148.

Stobart, J. (2010). A history of shopping: the missing link between retail and consumer revolutions. *Journal of Historical Research in Marketing*, 2, 342–349.

Strengers, Y. (2011). Negotiating everyday life: the role of energy and water consumption feedback. *Journal of Consumer Culture*, 11(3), 319–338.

Strengers, Y. (2013). *Smart Energy Technologies in Everyday Life: Smart Utopia?* New York: Springer.

Strengers, Y., & Hazas, M. (2019). Promoting smart homes. In J. Rinkinen, E. Shove & J. Torriti (eds.) *Energy Fables: Challenging Ideas in the Energy Sector.* London: Routledge. 78–87.

Strengers, Y., Maller, C., Nicolls, L. & Pink, S. (2014) Australia's rising air con use makes us hot and bothered. *The Conversation*. Available at: https://theconversation.com/australias-rising-air-con-use-makes-us-hot-and-bothered-20258, Accessed 20 September 2019.

Taylor, C. (1989). *Sources of the Self: The Making of the Modern Identity.* Harvard: Harvard University Press.

Tennøy, A., Gundersen, F., Hagan, O., Knapskog, M. & Uteng, T. (2017) Effects on traffic and emissions of densification in nodes in Bergen, Kristiansand and Oslo. TøI report 1575/2017. Available at: www.toi.no/publications/effects-on-traffic-and-emissions-of-densification-in-nodes-in-bergen-kristiansand-and-oslo-article34583-29.html, Accessed 20 September 2019.

TfNH (2018) Project Summary and Recommendations, Transport for New Homes, July. Available at: www.transportfornewhomes.org.uk/wp-content/uploads/2018/07/transport-for-new-homes-summary-web.pdf, Accessed 10 September 2019.

Thaler, R. & Sunstein, C. (2009). *Nudge: Improving Decisions about Health, Wealth and Happiness.* London: Penguin.

Thompson, C. (1996). Caring consumers: gendered consumption meanings and the juggling lifestyle. *Journal of Consumer Research*, 22, 388–407.

Torriti, J. (2015). *Peak Energy Demand and Demand Side Response.* London: Routledge.

Torriti, J. (2017). Understanding the timing of energy demand through time use data: time of the day dependence of social practices. *Energy Research & Social Science*, 25(C), 37–47.

Torriti, J. (2019). Elasticity. In J. Rinkinen, E. Shove & J. Torriti (eds.) *Energy Fables: Challenging Ideas in the Energy Sector.* Chapter 4. London: Routledge.

Trentmann, F. (2016). *Empire of Things: How We Became a World of Consumers, from the Fifteenth Century to the Twenty-first.* Milton Keynes: Penguin UK.

Trentmann, F., & Carlsson-Hyslop, A. (2018). The evolution of energy demand in Britain: politics, daily life, and public housing, 1920s–1970s. *The Historical Journal*, 61(3), 807–839.

Tukker, A. (2000). Life cycle assessment as a tool in environmental impact assessment. *Environmental Impact Assessment Review*, 20(4), 435–456.

Twilley, N. (2014). What do Chinese dumplings have to do with global warming. *New York Times Magazine*. Available at: www.nytimes.com/2014/07/27/maga zine/what-do-chinese-dumplings-have-to-do-with-global-warming.html, Accessed 9 January 2020.

UK Government. (2017a). The clean growth strategy. Available at: https://assets. publishing.service.gov.uk/government/uploads/system/uploads/attachment_ data/file/700496/clean-growth-strategy-correction-april-2018.pdf, Accessed 20 September 2019.

UK Government. (2017b). *The Clean Growth Strategy. Leading the way to a low carbon future*. Available at: www.gov.uk/government/publications/clean-growth-strat egy, Accessed 27 April 2018.

UKPIA (2019). Statistical Review 2019. Available at: www.ukpia.com/media-centre/news/2019/ukpia-2019-statistical-review/, Accessed 9 January 2020.

Urry, J. (2004). The 'system' of automobility. *Theory, Culture & Society*, 21(4–5), 25–39.

Verbeek, P. (2005). *What Things Do: Philosophical Reflections on Technology, Agency, and Design*. Pennsylvania, PA: The Pennsylvania State University Press.

Vörösmarty, C., Green, P., Salisbury, J. & Lammers, R. (2000). Global water resources: vulnerability from climate change and population growth. *Science*, 289 (5477), 284–288.

Wajcman, J. (2010). Feminist theories of technology. *Cambridge Journal of Economics*, 34(1), 143–152.

Walker, G., Simcock, N. & Day, R. (2016). Necessary energy uses and a minimum standard of living in the United Kingdom: energy justice or escalating expectations? *Energy Research & Social Science*, 18, 129–138.

Wallingford, H. (2015). CCRA2: updated projections of water availability for the UK, Final Report to the Committee on Climate Change. Available at: www. theccc.org.uk/wp-content/uploads/2015/09/CCRA-2-Updated-projections-of-water-availability-for-the-UK.pdf, Accessed 10 September 2019.

Warde, A. (2005). Consumption and theories of practice. *Journal of Consumer Culture*, 5(2), 131–153.

Warde, A. (2014). After taste: culture, consumption and theories of practice. *Journal of Consumer Culture*, 14(3), 279–303.

WEC, World energy council (2018). *World energy issues monitor 2018*. Available at: www.worldenergy.org/wp-content/uploads/2018/05/Issues-Monitor-2018-HQ-Final.pdf, Accessed 6 August 2019.

Wilhite, H., & Lutzenhiser, L. (1999). Social loading and sustainable consumption. *Advances in Consumer Research*, 26(1), 281–287.

Wilhite, H., Shove, E., Lutzenhiser, L. & Kempton, W. (2000). The legacy of twenty years of energy demand management: we know more about individual behaviour but next to nothing about demand. In E. Jochem, J. Sathaye & D. Bouille (eds.) *Society, Behaviour, and Climate Change Mitigation*. Dordrecht: Springer. 109–123.

Woolgar, S. (1991). The turn to technology in social studies of science. *Science, Technology, & Human Values*, 16(1), 20–50.

Yates, L., & Warde, A. (2015). The evolving content of meals in Great Britain. Results of a survey in 2012 in comparison with the 1950s. *Appetite*, 84, 299–308.

Zerubavel, E. (1979). *Patterns of Time in Hospital Life: A Sociological Perspective*. Chicago, IL: University of Chicago Press.

Zerubavel, E. (1985). *The Seven Day Circle: The History and Meaning of the Week*. London: Collier Macmillan.

Index

Printed in the United States
by Baker & Taylor Publisher Services

T0358280